Design of Experiments and their Implementations

Design of Experiments and their Implementations

Edited by **Justin Riggs**

LANRYE
INTERNATIONAL

New Jersey

Published by Clanrye International,
55 Van Reypen Street,
Jersey City, NJ 07306, USA
www.clanryeinternational.com

Design of Experiments and their Implementations
Edited by Justin Riggs

International Standard Book Number: 978-1-63240-137-3 (Hardback)

Printed in the United States of America.

Contents

Preface

I am honored to present to you this unique book which encompasses the most up-to-date data in the field. I was extremely pleased to get this opportunity of editing the work of experts from across the globe. I have also written papers in this field and researched the various aspects revolving around the progress of the discipline. I have tried to unify my knowledge along with that of stalwarts from every corner of the world, to produce a text which not only benefits the readers but also facilitates the growth of the field.

This book compiles authentic researches on advancements made within the field of design of experiments. It is a collection of analyzed scholarly contributions by various authors from across the globe. Each contribution is comprehensively presented in this book. The book is intended for scholars as well as specialists in the field.

Finally, I would like to thank all the contributing authors for their valuable time and contributions. This book would not have been possible without their efforts. I would also like to thank my friends and family for their constant support.

Editor

Robust Design and Taguchi Method Application

Helder Jose Celani de Souza, Messias Borges Silva,
Cinthia B. Moyses, Fernando Lopes Alberto,
Fabrício J. Pontes, Ubirajara R. Ferreira,
Roberto N. Duarte and
Carlos Eduardo Sanches da Silva

Additional information is available at the end of the chapter

1. Introduction

The objective of this chapter is to present an application of Taguchi Experimental Design Method. A healthcare case was chosen for this purpose. It is specifically applied to Molecular Assays in Clinical Laboratories and the main target is to determine the best parameters adjustment of a Molecular Assays Process in order to obtain the best diagnostic result for Venous Thromboembolism investigation [1].

1.1. Venous thromboembolism investigation by molecular assays process

Disorders of haemostatic mechanisms that predispose a person to thrombotic episodes are generally referred to as thrombophilia [2]. The incidence of one such disorder, Venous Thromboembolism, increases with age, ranging from 1 in 10,000 among children to 1 in 100 among elderly. Its most common clinical manifestations are deep vein thrombosis of the lower limbs and pulmonary embolism. Origins of thrombophilia risk factors can be both acquired and genetic, making the thromboembolic disease a complex, multifactorial trait [3].

The acquired risk factors include pregnancy, surgery, trauma, immobilization, advanced age, as well as previous episodes of thrombosis. The genetic risk factors most relevant are two mutations: the Factor V Leiden mutation (1691G>A) and the 20210G>A point mutation in the 3' untranslated region (UTR) of the Prothrombin gene. The first causes resistance to activated protein C and the second is associated with increased plasma factor II levels [4]. Both mutations act as gain of function, causing hypercoagulability.

There are two basic methods for investigating Factor V Leiden: functional (Activated Protein C Resistance) and molecular (such as PCR - Polymerase Chain Reaction - followed by restriction digestion, or real-time PCR). However, only the second method can be used to detect Prothrombin mutation (the functional methods are unacceptable due to overlapping between normal and carrier levels). Among molecular methods, real-time PCR offers many advantages over conventional PCR: higher sensitivity, quicker turn-around-time, better uniformity and objectivity in the analysis, and lower probability of contamination in the laboratory.

Real-time PCR with Fluorescence Resonance Energy Transfer (FRET) probes (also called hybridization probes) employ two labeled probes, commonly called FRET and anchor probes. These probes hybridize to the PCR product in head-to-tail fashion, at close proximity. Because the acceptor fluorophore emits light in a longer wavelength, signal detection is possible. One of these probes is labeled with a donor dye at the 3' end and the other is labeled with an acceptor dye at the 5' end. Quenched-FRET assays are similar to FRET assays, except in what they measure. Instead of measuring the increase in energy of the acceptor fluorophore, quenched-FRET assays measure, during amplification, the decrease in energy of the donor fluorophore. Use of a quencher molecule, such as black-hole quenchers, in place of an acceptor fluorophore enables multiplexing of more than one fluorophore. This permits the analysis of a greater number of mutations in the same tube. By performing a melting analysis after PCR to find the amplicon-probe melting temperatures, genotyping is achieved. Variation may be shown depending on the number of mismatches, the length of the mismatched duplex, the position of the mismatch and neighboring base pairs. Studies previously published have confirmed the efficiency of FRET real-time PCR for SNP detection and allelic discrimination of Factor V Leiden and Prothrombin (factor II) [5-10].

The FRET system requires four oligonucleotides. Therefore to conduct a successful experiment with reliable results two steps are necessary: careful design of the probes and primers and assays optimization.

1.2. Taguchi experimental design

The experimental design is widely used to optimize process parameter values in order to improve the quality properties of a product or a process. Full Factorial and One-Factor-at-the-time (OFTA) experiments are design methods that can possibly be used but requires a large number of experiments when the number of process parameters increases. Taguchi developed the foundations of Robust Design introduced in the 1950s and 1960s and the application of his method in electronics, automotive, photographic and many others industries has been an important factor in the rapid industrial growth of Japanese industries [11].

Among the various approaches to quality engineering products and processes, the method of Taguchi is identified for robust design [14]. The method deviates from the quality engineering concerns when it considers the objective to ensure good quality products and good process performance deliveries during the life cycle of these projects [12]. Taguchi methods are distinguished from other approaches to quality engineering by some specific concepts, as follow:

- Minimization of a quality loss function

- Maximization the signal to noise ratio

- Orthogonal Arrays

The strategy of experimental design used in the Taguchi method is based on orthogonal arrays and fractional factorial, in which not all possible combinations of factors and levels are tested. It is useful to estimate the effects of main factors on the process. The primary goal of this type of strategy is to obtain as much information about the effect of the parameters on the process with minimal experimental runs. In addition to the fact of requiring a smaller number of experiments, the orthogonal arrays still allow to test the factors using a mixing of number of levels.

Taguchi method uses a special design of Orthogonal Arrays that allows to study the whole parameter space with a limited number of experiments [12]. Besides, this method provides other advantages: it reduces economically the variability of the response variable, shows the best way to find out the optimum process conditions during laboratory experiments, it is an important tool for improving the productivity of the R&D activity and it can be applied to any process.

The usual steps to apply Taguchi experimental design [13] are: (a) to select the output variable(s) (response(s)) to be optimized; (b) to identify the factors (input variables) affecting output variable(s) and to choose the levels of these factors; (c) to select the appropriate Orthogonal Array; the arrays are found in literature [14]; (d) to assign factors and interactions to the columns of the array; (e) to perform experiments; at this step it is important to randomize the trials in order to minimize the systematic error; (f) to analyze the results using signal-to-noise ratio (S/N) analysis and analysis of variance (ANOVA); (g) to determine the optimal process parameters; (h) to perform confirmatory experiments, if it is necessary.

For the S/N ratio analysis, the appropriate S/N ratio function must be chosen: smaller-the-better, larger-the-better, nominal-the-better. The S/N ratio is a logarithmic function used to optimize the process or product design, minimizing the variability, as shown by Equation 1.

$$\eta = -10\log_{10}[\frac{1}{n}\sum_{i=1}^{n}\frac{1}{y_i^2}] \tag{1}$$

In the equation (1), η is the signal to noise ratio, y_i is the Quality Function Deviation, problem type "larger-the-better", which is the case of this application and, n corresponds the number of experiments runs.

The S/N ratio can be also understood as the inverse of variance and the maximization of S/N ratio allows reduction of the variability of the process against undesirable changes in neighbouring environment (also named uncontrollable factors or factors of noise). To minimize variability, the level of factor which produces the greatest value of S/N ratio must be chosen.

The analysis of variance (ANOVA) is applied in order to test the equality of several means, resulting in what process parameters (factors) are statistically significant.

In the methodology of experimental design, the test used to evaluate the significance of the levels changes of a factor or an interaction is a hypothesis test. In the case of full factorial, this test is an analysis of variance (ANOVA) [18]. When two levels of a factor generating have equal statistically mean responses, it is assumed that the factor does not affect the response of interest. When, instead, a significant difference is detected, the factor is important. For a full factorial with two factors A and B and two levels (+1, -1), the correspondent model can be shown in equation (2).

$$Y_{ijk} = \mu + A_i + B_j + AB_{ij} + \varepsilon_{ijk} \qquad (2)$$

Where:

i is the number of levels of the fator A;

j is the number of levels of the fator B and k, the number of replicas;

Y_{ijk} is the (ijk)th observation obtained in the experiment;

μ is the overall mean;

A_i is the effect of the ith treatment of Factor A;

B_j is the effect of the jth treatment of Factor B;

AB_{ij} is the effect of the ij-th AB interaction between factors;

ε_{ijk} is the component of random error.

The results of ANOVA are presented in a table that displays for each factor (or interaction) the values of the sum of squared (SS) deviations from the mean, the mean of squares (MS) and the ratio between the mean of squares effect and the mean of squares error (F). For background information many introductory texts on elementary statistical theory are available in literature and can also be found in most of the statistical packages for microcomputers [16-18].

In this chapter the Taguchi L27 Orthogonal Array employed for experimental design and data analysis, considers the search for the best conditions of operation, the effects of the main factors over the process, and the interactions among the factors. The Taguchi method was applied by Ballantyne et al. [15] for the optimization of conventional PCR assays using an L16 Orthogonal Array with four variables at two different levels each. The present research, however, is considered a more complex Taguchi's method application once it optimizes a process that uses real-time PCR using FRET probes with six three-levels factors.

2. Material and methods

Distinct DNA pools for Factor V Leiden and Prothrombin were employed. Each was consti-
tuted of samples from quality control programs whose genotypes were already known. The
mutation detection was performed by real-time PCR, followed by melting curve analysis with
adjacent fluorescent probes using the quenched-FRET principle.

With the exception of the reverse primers (Prothrombin: 5'- ATTACTGGCTCTTCCTGAGC3';
Factor V Leiden:5'TGCCCAGTGCTTAACAAGAC-3'), the primer and probe sequences for
both Prothrombin and Factor V Leiden genotyping were the same as those described by von
Ahsen et al. [6] and Ameziane et al. [10], respectively. The Factor V Leiden detection probe,
which was specific for the mutated allele, was 3'-labeled with 6-carboxyfluorescein (FAM).
The adjacent probe, which functioned as an anchor, was 5'-labeled with Cy5 and phosphorylated at its 3' end; this was to prevent probe elongation by *Taq* polymerase [5-8].

The Prothrombin detection probe, which was complementary to the wild-type allele, was 3'-
labeled with 6-carboxy-4',5'-dichloro-2',7'-dimethoxyfluorescein (JOE). The anchor probe was
5'-labeled with 6-carboxytetramethylrhodamine (TAMRA) and phosphorylated at its 3' end.
About 240 ng of genomic DNA in a final PCR volume of 25 µL were amplified and detected
in the Rotor-Gene 3000 (Corbett Research, Australia), as shown in Figure 1.

Sample Preparation Chapel   **Rotor Gene 3000**   **Sample Loading Carousel**

Figure 1. Experiments Environment: Sample Preparation Chapel and Rotor Gene 3000

For standardization, PCR reactions were performed using distinct Master Mixes: MMA,
Promega PCR Master Mix (1.5 mM MgCl$_2$); MMB, Promega PCR Master Mix (3.0 mM MgCl$_2$);
MMC, QIAGEN PCR Master Mix (1.5 mM MgCl$_2$), different concentrations of primers
(forward and reverse) and probes (FRET and anchor) and different PCR cycle numbers to test
for the best combination.

All six factors selected as "input variables" for standardization (Table 1) were investigated at
three different levels. These factors were selected for being essential components of a PCR.
Also, the levels tested cover the range suggested in the literature. Although a standard real-
time PCR usually uses a maximum of 45 to 50 cycles, some protocols, such as asymmetric PCR,
may require more.

Cycling and melting profiles were performed according to the following protocol: 95°C for 4 min as the initial denaturation step. This was followed by n cycles (n = 50, 65 or 85) of 95°C for 10 s, 53°C for 20 s and 72°C for 20 s. Thereafter, melting curve analysis of the duplex amplicon-probe was performed. Analysis started at 49°C and proceeded until to 88°C, at a linear rate of 1°C every 5 s.

Factor name	Column allocated to the factor in L27 Taguchi Orthogonal Array	Level 1	Level 2	Level 3	Unit
Master Mix	A	MMA	MMB	MMC	Composition/ supplier
Primer forward concentration (P1)	B	0.1	0.5	1.0	µM
Primer reverse concentration (P2)	E	0.1	0.5	1.0	µM
FRET probe concentration (S1)	J	0.2	0.3	0.4	µM
Anchor probe concentration (S2)	K	0.2	0.3	0.4	µM
Number of PCR cycles	L	45	65	85	cycles

Table 1. Experiments Factors and Levels

The Rotor-Gene software calculated the negative derivative of the fluorescence ratio with respect to temperature (-dF/dT), which was plotted against temperature (T°C). The melting curves were then converted to melting peaks. The "output variable" was the melting peak height measured after the melting analysis, which could be visualized in the negative derivative plot (-dF/dT x T°C). This variable was chosen as the output because it reflected the efficiency of the whole process, including both amplification and melting analysis.

PCR mix preparation, sample loading, and amplification reactions were carried out in three separate rooms. All rooms were subject to temperature and humidity control. In order to minimize the effect of pipetting errors between runs, amplification mix was prepared once and then divided into aliquots. These were stored, protected from light, at -20°C. Each aliquot was thawed only once at the time of use. In addition, in order to minimize the effect of inter-operator variation, a single person was responsible for the execution of the whole process. In order to maintain reagents and sample stability, each of the 27 experiments was performed over consecutive days in quadruplicate.

Tubes were randomly loaded in the 36 carousel rotor. These procedures, performed independently, were adopted for both Factor V Leiden and Prothrombin genotyping. The research equipment outputs fluorescence values at the origin point of the fluorescence curve. This constraint may lead to low output values for curves of high resolution and high values for curves of low resolution. This restriction was overcome by employing the derivative of the curve for Fluorescence and defining the output as the peak mean values or the peak values. Thus meaningful values for both Prothrombin genotyping and V Leiden were achieved.

The curve for Melting was generated with the use of Rotor-Gene 3000 equipment by using the curve for Fluorescence of each sample. In order to estimate the individual and interaction effects among the factors, a Taguchi L27 Orthogonal Array was employed, with four replicates for each experiment.

The Taguchi method differs from other quality engineering tools in terms of some specifics concepts, once it includes the minimization of the quality loss function, the maximization of the noise-to-signal ratio, a quadratic loss function [14], and the usage of Orthogonal Arrays [16-18]. The results were later compiled and analyzed through statistical methods using Statistica9 software.

3. Results

Tables 2 and 3 show the L27 Orthogonal Array results considering four replicates per run. Table 3 shows the experimental factors and the levels considered by Taguchi method and used to determine the optimal adjustments for the best final results.

The Factors A, B, E, J, K AND L are assigned to columns 1, 2, 5, 9, 10 and 11 in the L27. The Factors C, D, F, G, H, I, M and N are assigned to columns 3, 4, 6, 7, 8, 12, and 13 in L27 and represent the interaction effects being treated as dummy factors.

The output variability is reduced when the signal-to-noise ratio is maximized. In this Taguchi method application the design condition to reach this goal is larger is better.

In the sequence, Tables 2 and 3 are shown.

The same procedure is equally done for Factor V Leiden genotyping.

Figures 2 and 3 show the main effects for means and signal-to-noise ratio for Prothrombin genotyping, as follow.

3.1. Results from prothrombin genotyping analysis

By distributing the mean output values among experimental setups, it was possible to evaluate each experimental factor's impact on the output. In this case, the experimental factors major impact results in a descending order are E, B, A, H, L, K, G, J, F, N, M, C and D, varying from 60.29% to 1.83% respectively.

Statistical analysis was performed by applying Analysis of Variance (ANOVA) to the obtained results. Each ANOVA factor adopted two degrees of freedom, which corresponds to the number of levels adopted to a given experimental factor less one. Square sums of the two less

														Replicas			
Run	**A**	**B**	**C**	**D**	**E**	**F**	**G**	**H**	**J**	**K**	**L**	**M**	**N**	**1**	**2**	**3**	**4**
1	1	1	1	1	1	1	1	1	1	1	1	1	1	0.00001	0.00001	0.00001	0.00001
2	1	1	1	1	2	2	2	2	2	2	2	2	2	0.09000	0.08500	0.08500	0.07500
3	1	1	1	1	3	3	3	3	3	3	3	3	3	0.20000	0.16500	0.19000	0.18500
4	1	2	2	2	1	1	1	2	2	2	3	3	3	0.00001	0.00001	0.00001	0.00001
5	1	2	2	2	2	2	2	3	3	3	1	1	1	0.02000	0.00001	0.01500	0.00001
6	1	2	2	2	3	3	3	1	1	1	2	2	2	0.05500	0.04500	0.05000	0.05000
7	1	3	3	3	1	1	1	3	3	3	2	2	2	0.00001	0.00001	0.00001	0.00001
8	1	3	3	3	2	2	2	1	1	1	3	3	3	0.00500	0.00001	0.00001	0.00500
9	1	3	3	3	3	3	3	2	2	2	1	1	1	0.05500	0.06000	0.05000	0.06000
10	2	1	2	3	1	2	3	1	2	3	1	2	3	0.00001	0.00001	0.00001	0.00001
11	2	1	2	3	2	3	1	2	3	1	2	3	1	0.04500	0.00001	0.00001	0.00001
12	2	1	2	3	3	1	2	3	1	2	3	1	2	0.08000	0.06000	0.06000	0.08500
13	2	2	3	1	1	2	3	2	3	1	3	1	2	0.00001	0.00001	0.00001	0.00001
14	2	2	3	1	2	3	1	3	1	2	1	2	3	0.00500	0.00500	0.02000	0.00001
15	2	2	3	1	3	1	2	1	2	3	2	3	1	0.02000	0.00001	0.02000	0.01000
16	2	3	1	2	1	2	3	3	1	2	2	3	1	0.00001	0.00001	0.00001	0.00001
17	2	3	1	2	2	3	1	1	2	3	3	1	2	0.00001	0.00001	0.00001	0.00001
18	2	3	1	2	3	1	2	2	3	1	1	2	3	0.02000	0.02000	0.00001	0.02000
19	3	1	3	2	1	3	2	1	3	2	1	3	2	0.00001	0.00001	0.00001	0.00001
20	3	1	3	2	2	1	3	2	1	3	2	1	3	0.07000	0.06500	0.05500	0.06500
21	3	1	3	2	3	2	1	3	2	1	3	2	1	0.10000	0.09000	0.08500	0.09500
22	3	2	1	3	1	3	2	2	1	3	3	2	1	0.00001	0.00001	0.00001	0.00001
23	3	2	1	3	2	1	3	3	2	1	1	3	2	0.00001	0.02000	0.00001	0.02500
24	3	2	1	3	3	2	1	1	3	2	2	1	3	0.04500	0.03500	0.04000	0.03500
25	3	3	2	1	1	3	2	3	2	1	2	1	3	0.01000	0.00001	0.00001	0.00001
26	3	3	2	1	2	1	3	1	3	2	3	2	1	0.00001	0.00001	0.00001	0.00001
27	3	3	2	1	3	2	1	2	1	3	1	3	2	0.04000	0.03000	0.04000	0.02500

L27 Orthogonal Array and Replicates Results for Factor V Leiden

Table 2. Factor Prothrombin Gene L27 Orthogonal Array Results

influential factors were employed to estimate errors, due to the fact that a saturated Taguchi design was employed.

														Replicas			
Run	A	B	C	D	E	F	G	H	J	K	L	M	N	1	2	3	4
														L27 Orthogonal Array and Replicates Results for Factor V Leiden			
1	1	1	1	1	1	1	1	1	1	1	1	1	1	0.00001	0.00001	0.00001	0.00001
2	1	1	1	1	2	2	2	2	2	2	2	2	2	0.00001	0.00001	0.00001	0.00001
3	1	1	1	1	3	3	3	3	3	3	3	3	3	0.00001	0.00001	0.00001	0.00001
4	1	2	2	2	1	1	1	2	2	2	3	3	3	0.54000	0.46000	0.51500	0.48000
5	1	2	2	2	2	2	2	3	3	3	1	1	1	0.00001	0.00001	0.00001	0.00001
6	1	2	2	2	3	3	3	1	1	1	2	2	2	0.00001	0.00001	0.00001	0.00001
7	1	3	3	3	1	1	1	3	3	3	2	2	2	0.43000	0.43500	0.35500	0.39000
8	1	3	3	3	2	2	2	1	1	1	3	3	3	0.08000	0.10500	0.08500	0.07500
9	1	3	3	3	3	3	3	2	2	2	1	1	1	0.00001	0.00001	0.00001	0.00001
10	2	1	2	3	1	2	3	1	2	3	1	2	3	0.00001	0.00001	0.00001	0.00001
11	2	1	2	3	2	3	1	2	3	1	2	3	1	0.00001	0.00001	0.00001	0.00001
12	2	1	2	3	3	1	2	3	1	2	3	1	2	0.00001	0.00001	0.00001	0.00001
13	2	2	3	1	1	2	3	2	3	1	3	1	2	0.26000	0.29000	0.25000	0.24000
14	2	2	3	1	2	3	1	3	1	2	1	2	3	0.00001	0.00001	0.00001	0.00001
15	2	2	3	1	3	1	2	1	2	3	2	3	1	0.00001	0.00001	0.00001	0.00001
16	2	3	1	2	1	2	3	3	1	2	2	3	1	0.25500	0.25000	0.25500	0.24500
17	2	3	1	2	2	3	1	1	2	3	3	1	2	0.10000	0.10000	0.07500	0.06500
18	2	3	1	2	3	1	2	2	3	1	1	2	3	0.00001	0.00001	0.00001	0.00001
19	3	1	3	2	1	3	2	1	3	2	1	3	2	0.00001	0.00001	0.00001	0.00001
20	3	1	3	2	2	1	3	2	1	3	2	1	3	0.00001	0.00001	0.00001	0.00001
21	3	1	3	2	3	2	1	3	2	1	3	2	1	0.00001	0.00001	0.00001	0.00001
22	3	2	1	3	1	3	2	2	1	3	3	2	1	0.39500	0.43500	0.41000	0.42500
23	3	2	1	3	2	1	3	3	2	1	1	3	2	0.00001	0.00001	0.00001	0.00001
24	3	2	1	3	3	2	1	1	3	2	2	1	3	0.00001	0.00001	0.00001	0.00001
25	3	3	2	1	1	3	2	3	2	1	2	1	3	0.55000	0.53500	0.53500	0.58000
26	3	3	2	1	2	1	3	1	3	2	3	2	1	0.15500	0.18000	0.14000	0.12500
27	3	3	2	1	3	2	1	2	1	3	1	3	2	0.00001	0.00001	0.00001	0.00001

Table 3. Factor V Leiden L27 Orthogonal Array Results

Therefore, error term was considered to have four degrees of freedom. Table 4 shows ANOVA results for output average value in experiments involving Prothrombin genotyping. The ANOVA analysis revealed which experimental factors were significant to the process output.

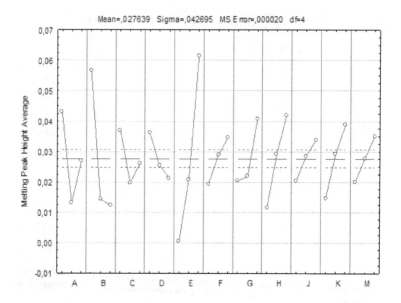

Figure 2. Means Main Effects for Prothrombin Genotyping Experiments

The decreasing order of significance for maximization of output for Prothrombin genotyping is as follows: E (concentration of primer reverse); B (concentration of primer forward); H, A (Master Mix), L (Number of PCR cycles), G, C, F, D, N and K (Anchor probe concentration).

By observing the interaction diagram of the L27 Taguchi method employed, it is evident that columns H, G, C, F, D and N contain information about interaction between physical factors. This suggests that interactions between the levels of physical factors significantly influence the outcome of the process. The levels of factors that will maximize the output for Prothrombin genotyping are displayed in the Table 5.

In order to estimate experimental conditions that maximize the robustness of the process, the ANOVA test was carried out on experimental results obtained for signal-to-noise ratio, as shown in Table 6.

For the same reasons presented before, square sums of the two less influential factors for robustness, C and D, were employed to estimate errors. Once more the number of degrees of the error was set to four, which corresponds to the sum of degrees of freedom of the factor employed to estimate the error term.

The ANOVA analysis also revealed the order of significance for maximizing the process's robustness. The decreasing order of significance is as follows: E (concentration of primer reverse); B (concentration of primer forward); A (Master Mix), H, L (Number of PCR cycles),

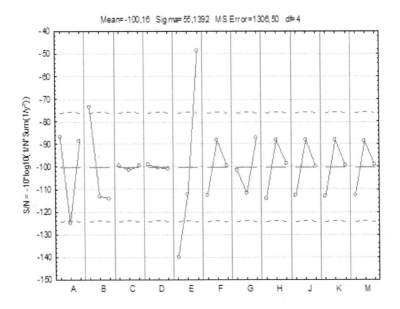

Figure 3. Signal-To-Noise Ratio Main Effects for Prothrombin Genotyping Experiments

K (Anchor probe concentration), G, J (FRET probe concentration), N, F and M. By observing the interaction diagram of the L27 Taguchi method employed, it's apparent that columns H, G, N, F and M contain information about interaction between physical factors. This suggests that interactions play a significant role in process robustness.

The levels of factors that will maximize robustness are also displayed in Table 5. For the factors Master Mix, Primer Forward Concentration, and Primer Reverse Concentration, let us compare their recommended levels for output optimization and robustness maximization. Levels that maximize Prothrombin genotyping output are the same as those that maximize process robustness. The remaining factors (Anchor probe concentration and number of PCR cycles) have distinct levels maximize output and robustness.

3.2. Results from factor V leiden analysis

Analysis performed on results obtained for Prothrombin genotyping were integrally repeated on results obtained for Factor V Leiden genotyping and the recommended values are shown in Table 7.

In this case, the experimental factors major impact results in descending order were E, B, L, H, A, F, C, N, J, M, D, G and K, varying from 61.49% to 0.41% respectively.

Analysis of Variance on Melting Peak Height Average						
Mean =0 .027639			Sigma =0 .042695			
Factors (S – Source)	SS Sum of Squares	df Degree of Freedom	MS Mean Square	F F-Ratio	P P-Value	Result
A	0.004019	2	0.002010	101.5351	0.000	Significant
B	0.011178	2	0.005589	282.3947	0.000	Significant
C	0.001388	2	0.000694	35.0614	0.003	Significant
D	0.001052	2	0.000526	26.5877	0.005	Significant
E	0.017559	2	0.008779	443.5877	0.000	Significant
F	0.001091	2	0.000545	27.5526	0.005	Significant
G	0.002310	2	0.001155	58.3509	0.001	Significant
H	0.004239	2	0.002119	107.0877	0.000	Significant
K	0.000796	2	0.000398	20.1140	0.008	Significant
L	0.002690	2	0.001345	67.9561	0.001	Significant
N	0.000994	2	0.000497	25.1140	0.005	Significant
Residual Error	0.000079	4	0.000020			

Table 4. ANOVA table for output values for Prothrombin Genotyping Experiments

The maximization of output for coagulation Factor V Leiden, in descending order of significance, is as follows: E (Primer reverse Concentration); B (Primer forward Concentration) and L (Number of PCR cycles), H, A (Master Mix), F, C, G, N, J (FRET probe concentration).

The interaction diagram of the L27 Taguchi method reveals that columns H, F, C, G and N contain information about interaction between physical factors. This suggests that interactions between the levels of physical factors significantly influence the outcome of the process [11].

3.3. Confirmation experiments

Confirmation experiments were conducted using six samples, each of them in two replicates and are shown in the table 8. Also included were the factors adjusted to the recommended levels for process optimization and control samples.

Factor Index	Factor Name	Taguchi Factor Levels recommended for output optmization	Physical Value linked to Factor Level	Taguchi Factor Levels recommended for robustness maximization	Physical Value linked to Factor Level	Unit
A	Master Mix	1	MMA	1	MMA	Composition /supplier
B	Primer forward Concentration (P1)	1	0.1	1	0.1	µM
E	Primer reverse Concentration (P2)	3	1	3	1	µM
J	FRET probe concentration (S1)	Not significant	Not significant	1	0.2	µM
K	Anchor probe concentration (S2)	3	0.4	2	0.3	µM
L	Number of PCR cycles	3	85	2	65	cycles

Table 5. Factor levels recommended for output optimization or robustness maximization for Prothrombin Genotyping Experiments

The expected results are positive values and as much higher as possible to be considered good results for clinical significance and diagnostics qualitative analysis. The reproducibility is another important result for this application and the lowest standard deviation among replicates is also desirable which is also shown in the table 8. The results disclosed significant "responses," with values above zero. Such a finding clearly demonstrates that the recommendations about the conditions for the best process adjustments obtained by the Taguchi method meet the requirements and goals of this study.

A significant advantage to using the Taguchi method is the time and cost saved. Using the standard factorial design (or a non-formal method), will produce a much higher number of assays than will a fractional factorial as the Taguchi method uses. Extensively used to optimize engineering processes, the method incorporates one primary experiment to study the main effects of each factor, modeling some of the important interactions. Secondary Taguchi arrays can then be designed from the primary results, to narrow the optimal windows for each factor.

The method's strength lies in its Orthogonal Array design; each level of each factor occurs in an equal number of times across the entire array. Its potential savings are apparent when compared to factorial design. With the Taguchi method only 27 experiments were needed. For the same number of factors and levels examined, full factorial design requires 729 experiments [15, 16, 18].

		Analysis of Variance on S/N Ratio (Larger- the – Better)				
	Mean = -100.16			Sigma = 55.1392		
Factors (S - Source)	SS Sum of Squares	df Degree of Freedom	MS Mean Square	F F-Ratio	P P-Value	Result
A	8271.62	2	4135.81	406.00	0.000	Significant
B	9470.42	2	4735.21	464.84	0.000	Significant
E	39531.31	2	19765.66	1940.33	0.000	Significant
F	2633.31	2	1316.65	129.25	0.000	Significant
G	2674.61	2	1337.31	131.28	0.000	Significant
H	3034.27	2	1517.13	148.93	0.000	Significant
J	2655.79	2	1327.89	130.35	0.000	Significant
K	2688.69	2	1344.34	131.97	0.000	Significant
L	2828.24	2	1414.12	138.82	0.000	Significant
M	2570.21	2	1285.10	126.15	0.000	Significant
N	2649.50	2	1324.75	130.05	0.000	Significant
Residual Error	40.75	4	10.19			

Table 6. ANOVA results for Signal-To-Noise ratios for Prothrombin Genotyping Experiments

This work suggests new studies of similar processes employing the full factorial DOE technique [16] or RSM - Response Surface Method [20] using the Taguchi method. Such studies should yield a better, more accurate estimation of the significance of experimental factors and interactions among factor levels on process outputs. It must be emphasized that the results obtained in this research should not be extrapolated to other clinical processes.

4. Conclusions

The most relevant genetic risk factors associated with thrombophilia are the Factor V Leiden and the 20210G>A point mutation in the Prothrombin gene. Employing the Taguchi method, the study successfully optimized a screening method for Prothrombin genotyping and for Factor V Leiden mutation detection. The Taguchi Experimental Design method [16] proved an efficient tool in determining the different levels that maximize the process output. Here that is defined as the melting peaks at the derivative plot (-dF/dT x T°C) and the relevance of factors obtained by Taguchi method.

Regarding Prothrombin genotyping, data analysis uncovered the most significant factors for maximizing the process's output and robustness. Those factors, in decreasing order, are: Primer Reverse Concentration, Primer Forward Concentration, Master Mix Type, and number of cycles in PCR. In addition, the adjustment levels for maximization of the process output are:

Factor Index	Factor Name	Taguchi Factor Levels recommended for output optmization	Physical Value linked to Factor Level	Taguchi Factor Levels recommended for robustness maximization	Physical Value linked to Factor Level	Unit
A	Master Mix	3	MMC	3	MMC	Composition /supplier
B	Primer forward Concentration (P1)	3	1	3	1	µM
E	Primer reverse Concentration (P2)	1	0.1	1	0.1	µM
J	FRET probe concentration (S1)	2	0.3	Not Significant	Not Significant	µM
L	Number of PCR cycles	3	85	3	85	cycles

Table 7. Factor levels recommended for output optimization or robustness maximization for Factor V Leiden Genotyping Experiments

primer reverse concentration = 1 µM, primer forward concentration = 0.1 µM, master mix type MMA (Promega PCR Master Mix) and number of PCR cycles = 85 cycles. FRET probe concentration was considered non-significant.

Data analysis for Factor V Leiden genotyping uncovered the most significant factors for maximizing the process's output. Those factors, in decreasing order, are: Primer Reverse Concentration, Primer Forward Concentration, Number of Cycles in the reaction, Master Mix Type and FRET probe concentration. The adjustment levels that lead to maximizing process output are: Primer Reverse Concentration = 0.1 µM, Primer Forward Concentration = 1 µM, number of cycles = 85, Master Mix Type MMC and FRET probe concentration equal to 0.2 µM. Anchor probe concentration was considered as non-significant.

The same kind of analysis was performed for process robustness for both Prothrombin genotyping and Factor V Leiden. The proper levels, as well as the recommended levels of the experimental factors, for maximizing robustness were pointed out and revealed as significant.

Analysis also showed that interactions among factor levels play a significant role in maximizing both process output and robustness for both Prothrombin genotyping and Factor V Leiden. The nature and intensity of this interaction should be further investigated. One other thing is worth mentioning. All reactions for Factor V Leiden and Prothrombin genotyping described in this paper were performed independently. However, multiplexing (simultaneously conducting both reactions in the same tube) may be enabled through the use of different fluorophores (FAM and JOE) for Factor V and Prothrombin probe labeling. This permits the operator to load more samples in the equipment. Usually, different conditions of an assay

Run	Prothrombin (Factor II)	Standard Deviation	Fator V Leiden	Standard Deviation
1	0.350		0.480	
1	0.330	0.014	0.450	0.021
2	0.280		0.550	
2	0.320	0.028	0.690	0.099
3	0.195		0.460	
3	0.185	0.007	0.500	0.028
4	0.155		x	
4	0.190	0.025	x	x
5	x		0.600	
5	x	x	0.615	0.011
6	0.320		0.760	
6	0.320	0.000	0.830	0.049
am ctrl FII	0.150		x	
am ctrl FII	0.175	0.018	x	x
am ctrl FV	x		0.580	
am ctrl FV	x	x	0.585	0.004

Table 8. Confirmation Results for Prothrombin and Factor V Leiden using the Taguchi recommended Factors Levels

optimization are tested independently, in different assays, in which all conditions are kept constant except for the one being tested.

This process is usually costly and time-consuming. Neither does it assure that all possible combinations are tested. Therefore, the Taguchi method offers several advantages over the traditional optimization process; it allows for the performing different experiments, testing different levels for each factor in just a few days. This saves and reduces the time needed for the complete optimization and standardization processes of new diagnostic tests. In addition, it provides a mathematical support not only for the choice of the condition that generates the best result but also for the determination of the significant factors in the reaction. Such a benefit allows researchers to eliminate the non-relevant experimental factors for posterior fine-scale adjustments using the full factorial DOE technique [16] or Response Surface Methodology [19], when necessary.

The DOE methods can also be applied to a variety of quantitative tests. In conclusion, in clinical analysis laboratories that develop in house diagnostic tests, especially in the R&D area, applying the Taguchi method is an alternative and efficient approach for fast, low-cost assays optimization. It is important to clarify that Taguchi Method will not supply the final diagnostics for the patient but allows the best and optimal clinical assays factors adjustments to support a subsequent qualitative analysis by the clinical technician.

Acknowledgements

The authors are grateful to CAPES, Process PE024/2008 (Pro-Engineering Program - a Brazilian Government Program) and to Fleury Diagnostics.

Author details

Helder Jose Celani de Souza[1*], Messias Borges Silva[2,3], Cinthia B. Moyses[4], Fernando Lopes Alberto[4], Fabrício J. Pontes[3], Ubirajara R. Ferreira[3], Roberto N. Duarte[5] and Carlos Eduardo Sanches da Silva[6]

*Address all correspondence to: hcelani@uol.com.br

1 Siemens Healthcare Diagnostics, Sao Paulo State University – UNESP, Production Engineering Department, Guaratingueta, SP, Brazil

2 São Paulo University – USP, Lorena, SP, Brazil

3 Sao Paulo State University, UNESP, Guaratingueta, SP, Brazil

4 Fleury Diagnostics, São Paulo, SP, Brazil

5 CEFET - São Joao da Boa Vista, SP, Brazil

6 Federal University of Itajuba, Itajuba, MG, Brazil

References

[1] Molecular assay optimized by Taguchi experimental design method for venous thromboembolism investigation, Molecular and Cellular Probes Journal, 2011, 25: 231-237.

[2] British Committee for Standards in Haematology, Guidelines on investigation and management of thrombophilia. J. Clin Pathol 1990; 43: 703-10.

[3] Rosendaal F.R., Venous thrombosis: a multicausal disease. Lancet 1999; 353: 1167-73.

[4] Franco R.F., Reitsma P.H., Genetic risk factors of venous thrombosis. Hum Genet 2001; 109: 369-84.

[5] Lay M.J., Wittwer C.T., Real-time fluorescence genotyping of factor V Leiden during rapid-cycle PCR. Clin Chem 1997; 43: 2262-7.

[6]   von Ahsen N., Schütz E, Armstrong V.W., Oellerich M., Rapid detection of pro-
      thrombotic mutations of Prothrombin (G20210A), Factor V Leiden (G1691A), and
      methylenetetrahydrofolate reductase (C677T) by real-time fluorescence PCR with the
      LightCycler. Clin Chem 1999; 45: 694-6.

[7]   Neoh S.H., Brisco M.J., Firgaira F.A., Trainor K.J., Turner D.R., Morley A.A., Rapid
      detection of the factor V Leiden (1691 G>A) and haemochromatosis (845 G>A) muta-
      tion by fluorescence resonance energy transfer (FRET) and real time PCR. J Clin Path-
      ol 1999; 52: 766-9.

[8]   Sevall J.S., Factor V Leiden genotyping using real-time fluorescent polymerase chain
      reaction. Mol Cell Probes 2000; 14: 249–53.

[9]   Parks S.B., Popovich B.W., Press R.D., Real-time polymerase chain reaction with fluo-
      rescent hybridization probes for the detection of prevalent mutations causing com-
      mon thrombophilic and iron overload phenotypes. Am J Clin Pathol 2001; 115:
      439-47.

[10] Ameziane N., Lamotte M., Lamoril J., Lebret D., Deybach J.C., Kaiser T., de Prost D.,
      Combined Factor V Leiden (G1691A) and Prothrombin (G20210A) genotyping by
      multiplex real-time polymerase chain reaction using fluorescent resonance energy
      transfer hybridization probes on the Rotor-Gene 2000. Blood Coagul Fibrinolysis
      2003; 14: 421-4.

[11] Padke M.S., Quality Engineering Using Robust Design. Prentice Hall, Englewood
      Cliffs, NJ; 1989.

[12] Ross P.J., Taguchi Techniques for Quality Engineering. McGraw-Hill, USA, 43–73;
      1996.

[13] Barrado E., Vega M., Grande P., Del Valle J.L., Optimization of a purification ethod
      for metal-containing wastewater by use of a Taguchi experimental design. 1996; 30:
      2309–2314.

[14] Taguchi, G., Konishi, S., Taguchi Methods: Orthogonal Arrays and Linear Graphs.
      American Supplier Institute, USA; 1987.

[15] Ballantyne K.N., van Oorschot R.A., Mitchell R.J., Reduce optimisation time and ef-
      fort: Taguchi experimental design methods. Forensic Sci Int: Genetics 2008; Suppl
      Series 1: 7-8.

[16] Montgomery D.C., Design and Analysis of Experiments. NY, John Wiley & Sons;
      2003.

[17] Taguchi G., Elsayed E.A., Hsiang T., Quality Engineering in the Production Systems,
      McGraw-Hill; 1987.

[18] Montgomery D.C., Runger G.C., Applied Statistics and Probability for Engineers.
      New York: John Wiley & Sons; 2007.

[19] Vining, G.G., Myers, R.H., Combining Taguchi and response surface philosophies: a dual response approach. Journal of Quality Technology 22: 38-45, 1990.

Chapter 2

On the Effect of Fabrication and Testing Uncertainties in Structural Health Monitoring

H. Teimouri, A. S. Milani and R. Seethaler

Additional information is available at the end of the chapter

1. Introduction

Sensitive engineering structures are designed to be safe such that catastrophic failures can be avoided. Traditionally, this has been achieved by introducing safety factors to compensate for the lack of considering a structure's full-scale behavior beyond the expected loads. Safety factors create a margin between real-time operational loading and residual strength remaining in the structure. Historically, although fail-safe and safe-life methodologies were among design strategies for many years, the increasing impact of economical considerations and emerging inspection technologies led to a new design strategy called damage tolerance strategy [1]. Damage tolerant designed structures have an added cost which is related to the frequency and duration of inspections. For such structures, inspection intervals and damage thresholds are estimated and at every inspection the structure's health is investigated by looking for a maximum flaw, crack length and orientation. If necessary, modified investigation times are proposed, especially at vulnerable locations of the structure. Other limitations of the damage tolerant strategy include a lack of continuous assessment of the structure's health status and the need to pause the regular operation of the structure during off-line inspections. Over time, beside some historical catastrophic failures, the advancement of nondestructive technologies and economical benefits have directed designers to the introduction of the concept of Structural Health Monitoring (SHM). It may be hard to find a comprehensive and consistent definition for SHM, but as Boller suggested in [1], "SHM is the integration of sensing and possibly actuation devices to allow the loading and damage state of the structure to be monitored, recorded, analyzed, localized, quantified and predicted in a way that nondestructive testing becomes an integral part of the structure". This definition contains two major elements: load monitoring and damage diagnosis as the consequence of operational loading (which is often subject to a stochastic nature).

The review of literature shows an increasing number of research programs devoted to the development of damage identification systems to address problems such as assuming cost-effective methods for optimal numbering and positioning of sensors; identification of features of structures that are sensitive to small damage levels; the ability to discriminate changes caused by damage from those due to the change of environmental and testing conditions; clustering and classification algorithms for discrimination of damaged and undamaged states; and comparative studies on different damage identification methods applied to common datasets [2]. These topics are currently the focus of various groups in major industries including aeronautical [3, 4], civil infrastructure [5], oil [6, 7], railways [8], condition monitoring of machinery [9, 10], automotive and semiconductor manufacturing [2]. In particular, new multi-disciplinary approaches are increasingly developed and used to advance the capabilities of current SHM techniques.

2. Motivation of this study

A standard SHM technique for a given structure compares its damaged and healthy behaviors (by contrasting signals extracted from sensors embedded at specific points of the structure) to the database pre-trained from simulating/testing the behavior of the structure under different damage scenarios. Ideally, the change in the vibration spectra/stress-strain patterns an be related to damage induced in the structure, but it is possible at the same time that these deviations from a healthy pattern are caused by imperfect manufacturing processes including uncertainty in material properties or misalignment of fibers inside the matrix (in the case of composite structures), an offset of an external loading applied to the structure during testing, etc. Based on a strained-based SHM, this article addresses the important effect of manufac-turing/testing uncertainties on the reliability of damage predictions. To this end, as a case study a benchmark problem from the literature is used along with a finite element analysis and design of experiments (DOE) method. Among several existing DOE experimental designs (e.g., [11-16]) here we use the well-known full factorial design (FFD).

3. Case study description

The structure under investigation is a composite T-joint introduced in [17], where a strain-based structural health monitoring program, GNAISPIN (Global Neural network Algorithm for Sequential Processing of Internal sub Networks), was developed using MATLAB and NASTRAN-PATRAN. The T-joint structure, shown in Figure 1, consists of four major segments including the bulkhead, hull, over-laminates and the filler section. The finite element model of the structure is assumed to be two-dimensional (2D) and strain patterns are considered to be identical in the thickness direction of the structure. The geometrical constraints and applied load are also shown in Figure 1. The left-hand side constraint only permits rotation about the z-axis and prevents all other rotational and translational degrees of freedom. The right-hand side constraint permits translation along the x-axis (horizontal direction) and rotation about the z-axis. The displacement constraints are positioned 120mm away from the corresponding edges of the hull. The structure is subjected to a pull-off force of 5 kN. In [17], several delaminations were embedded in different locations of the structure, but in this study only a single delamination case is considered between hull and the left overlaminate. The strain distribution is then obtained for nodes along the bond-line (the top line of the hull between the right- and left-hand constraints), which are the nodes most affected by the presence of embedded delamination.

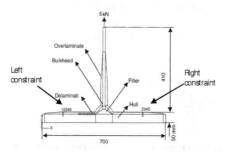

Figure 1. Geometry of the T-joint considered in the case study [17]

Using ABAQUS software, two dimensional orthotropic elements were used to mesh surfaces of the bulkhead, hull, and overlaminates, whereas isotropic elements were used to model the filler section. The elastic properties of the hull, bulkhead, and the overlaminates [17] correspond to 800 grams-per-square of plain weave E-glass fabric in a vinylester resin matrix (Dow Derakane 411-350). The properties of the filler corresponded to chopped glass fibers in the same vinylester resin matrix as summarized in Table 1.

Elastic Properties	Hull and Bulkhead	Overlaminate	Filler (quais-isotropic)
E1 (GPa)	26.1	23.5	2.0
E2 (GPa)	3.0	3.0	
E3 (GPa)	24.1	19.5	
v12=v23	0.17	0.17	0.3
v13	0.10	0.14	
G12=G23 (GPa)	1.5	1.5	0.8
G13 (GPa)	3.3	2.9	

Table 1. Elastic properties of the T-joint components

In order to verify the developed base ABAQUS model, strain distributions along the bond-line for the two cases of healthy structure and that with an embedded delamination are compared to the corresponding distributions presented in [17]. Figures 2.a and 2.b show a good accordance between the current simulation model and the one presented in [17] using NASTRAN-PATRAN. The only significant difference between the two models is found at the middle of the T-joint where results in [17] show a significant strain drop compared to the ABAQUS simulation. Figure 3 also illustrates the 2D strain distribution obtained by the ABAQUS model for the healthy structure case.

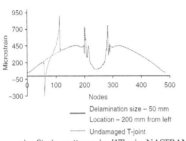

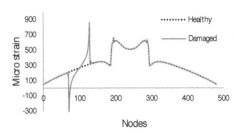

a) Strain patterns in [17] via NASTRAN-PATRAN model

b) The strain pattern obtained via ABAQUS model (the delamination size and location were identical to the NASTRAN-PATRAN model)

Figure 2. Comparison of strain distributions along the bond-line of the T-joint for different cases

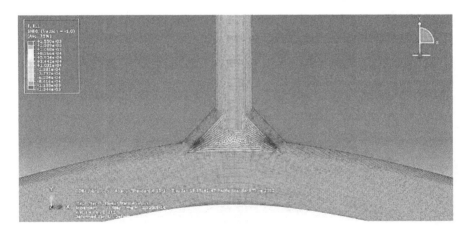

Figure 3. Strain field in the healthy T-joint via ABAQUS model (notice the symmetrical pattern)

Next, using the ABAQUS model for the DOE study, fiber orientations in the bulkhead, hull and overlaminate as well as the pull-off loading offset were considered as four main factors via a full factorial design, which resulted in sixteen runs for each of the health states (healthy and damaged structure). Two levels for each factor were considered: 0 or +5 degrees counter-clockwise with respect to the x-axis (Figure 4). Table 2 shows the assignment of considered factors and their corresponding levels. Table 3 represents the full factorial design for the two structural health cases.

	Factors	Coding	Levels (in degrees)
	Overlaminate	A	0 or 5
Regions of fiber angle error (misalignment)	Bulkhead	B	0 or 5
	Hull	C	0 or 5
Loading offset	Loading angle	D	0 or 5

Table 2. Factors and the corresponding levels considered in the DOE study

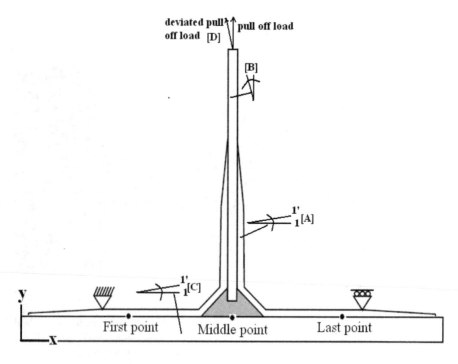

Figure 4. Schematic of study factors along with the position of the first, middle and the last nodes considered during the first DOE analysis

	Factors (all angles in degrees)				
run	A	B	C	D	Case
1	0	0	0	0	
2	5	0	0	0	
3	0	0	5	0	
4	5	0	5	0	
5	0	5	5	0	
6	5	5	5	0	
7	0	5	0	5	No Delamination
8	5	5	0	5	
9	0	0	0	5	
10	5	0	0	5	
11	0	0	5	5	
12	5	0	5	5	
13	0	5	5	5	
14	5	5	5	5	
15	0	5	0	0	
16	5	5	0	0	
17	0	0	0	0	
18	5	0	0	0	
19	0	0	5	0	Delamination of 50mm long at 200mm long from left edge
20	5	0	5	0	
21	0	5	5	0	
22	5	5	5	0	
23	0	5	0	0	
24	5	5	0	0	
25	0	0	0	5	
26	5	0	0	5	
27	0	0	5	5	
28	5	0	5	5	
29	0	5	5	5	
30	5	5	5	5	
31	0	5	0	5	
32	5	5	0	5	

Table 3. Full factorial design resulting in a total of 32 simulations (2^4 for the healthy structure and 2^4 for the damaged structure)

In order to illustrate the importance of the effect of uncertainty in fiber misalignment (e.g., during manufacturing of the structure's components), one can readily compare the difference between the strain distributions obtained for a case containing, e.g., 5° misalignment in the overlaminate (i.e., run # 2 in Table 3) and that for the perfectly manufactured healthy case (run # 1). A similar difference can be plotted between the case without any misalignment but in the presence of delamination (damage)-- which corresponds to run # 17 – and the perfectly manufactured healthy case (run # 1). These differences are shown in Figure 5.

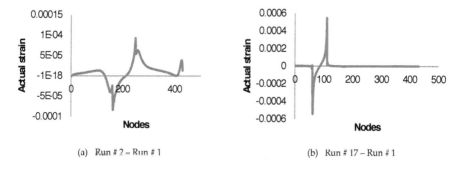

      (a) Run # 2 – Run # 1   (b) Run # 17 – Run # 1

Figure 5. Differences in strain distributions of sample runs in Table 3

By comparing the strain distributions in Figures 5.a and 5.b one can conclude that 5 degrees misalignment of fibers in the overlaminate (run # 2) has resulted in a significant deviation from the base model (run # 1) compared to the same deviations caused by the presence of delamination (run # 17); and hence, emphasizing the importance of considering fiber misalignment in real SHM applications and database developments. The next section is dedicated to perform a more detailed factorial analysis of results and obtain relative effects of the four alignment factors A, B, C, and D as samples of uncertainty sources in practice.

4. DOE effects analysis

Two different approaches are considered in the effects analysis; a point-to-point and an integral analysis. In the point-to-point approach, the difference between the horizontal strain values at three locations along the bond-line (first, middle and the last node in Figure 4) and those of the ideal case are considered as three output variables. On the other hand, the integral approach continuously evaluates the strain along the bond line where the number of considered points (sensors) tends to infinity. In fact the strain values obtained from the FE analysis would correspond to the strain data extracted from sensors embedded in the T-joint. The integral analysis for each given run, calculates the area under the strain distribution along the bond line, minus the similar area in the ideal case. The comparison of the two approaches, hence, provides an opportunity to assess the impact of increasing the number of sensors on the performance of SHM in the presence of manufacturing errors (here misalignments). For each approach, the most dominant factors are identified via comparing their relative percentage contributions on the output variables as well as the corresponding half-normal probability plots (see [16] for more theoretical details). Subsequently, ANOVA analysis was performed to statistically determine the significance (F-value) of key factors.

4.1. Point-to-point analysis results

Figure 4 shows the position of nodes assigned for the point-to-point analysis strategy. The first and last sensor points are considered to be 50mm away from the nearest constraint on the contact surface of hull and overlaminate. The middle point is located below the pull-off load point. Table 4 shows the results of FE runs based on the factor combinations introduced in Table 3. As addressed before, the presented data for the first group of runs (i.e., for healthy structures – runs 1 to 16) are the difference between strain values of each run and run 1; while the corresponding data for the second group (damaged T-joint – runs 17 to 32) represent the difference between strain values for each run and run 17. Table 5 represents the ensuing percentage contributions of factors and their interactions at each node for the two cases of healthy and delaminated T-joint. For the first node, which is close to the most rigid constraint on the left hand side of the structure, the only important factors are the misalignment of fibers in the hull (factor C) and its interaction with the loading angle offset (CD). This would be explained by the type of constraints imposed on the structure which is free horizontal translation of the opposite constraint on the right side. Figure 6 shows the half normal probability plot of the factor effects for the 1st node, confirming that factors C and CD are distinctly dominant parameters affecting the strain response at this node.

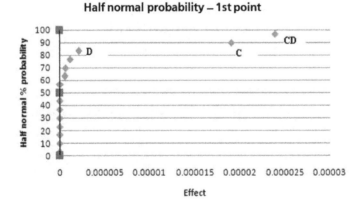

Figure 6. Half normal probability plot using the response at the 1st node during point-to-point analysis (for healthy structure)

Logically, one would expect that the mid node response would be strongly influenced by any loading angle offset as it can produce a horizontal force component and magnify the effect of the free translation boundary condition on the neighboring constraint; therefore, for the middle point response, the misalignment of fibers in the hull (C) and the loading angle error (D) and their interactions (CD) are the most significant factors, as also shown from the corresponding half normal probability plot in Figure 7. Finally, due to the short distance of the last (3rd) measuring node to the right constraint point and the strong influence of the large hull section beneath this measuring node, the parameter C was found to be the most dominant factor, followed by D, CD, AC, AD, and ACD (Figure 8).

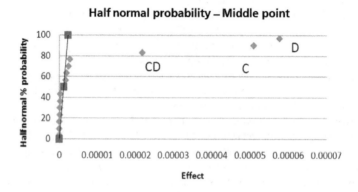

Figure 7. Half normal probability plot using the response at the 2nd node during the point-to-point analysis (for healthy structure)

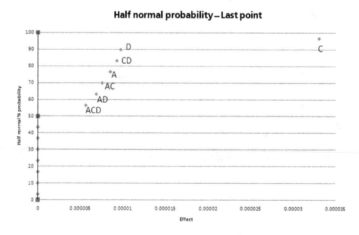

Figure 8. Half normal probability plot using the response at the 3rd sensor point during point-to-point analysis (for healthy structure)

Run	Factors				Response			Structure's health status
	A	B	C	D	@1st node	@middle node	@end node	
1	0	0	0	0	0	0	0	
2	5	0	0	0	1.0979E-06	2.273E-06	0.000013536	
3	0	0	5	0	4.23422E-05	3.0625E-05	4.04549E-05	
4	5	0	5	0	4.4637E-05	0.000029522	5.78129E-05	
5	0	5	5	0	4.23426E-05	0.000032361	4.04549E-05	
6	5	5	5	0	4.46377E-05	3.1389E-05	5.78129E-05	No Delamination (Healthy)
7	0	5	0	5	2.58877E-05	0.000040385	1.21364E-05	
8	5	5	0	5	2.71377E-05	0.000038519	5.23E-07	
9	0	0	0	5	2.58911E-05	0.000036949	0.000012135	
10	5	0	0	5	2.71419E-05	0.000034988	5.219E-07	
11	0	0	5	5	2.17078E-05	0.000111352	2.26219E-05	
12	5	0	5	5	2.16311E-05	0.000105259	3.74989E-05	
13	0	5	5	5	2.17102E-05	0.000115452	2.26229E-05	
14	5	5	5	5	0.000021634	0.000109486	3.74999E-05	
15	0	5	0	0	5E-10	1.445E-06	0	
16	5	5	0	0	1.0987E-06	3.83E-06	1.35358E-05	
	A	B	C	D	@1st node	@middle node	@end node	
17	0	0	0	0	0	0	0	
18	5	0	0	0	8.428E-07	1.744E-06	1.35359E-05	
19	0	0	5	0	4.22965E-05	0.000030471	4.04562E-05	Delamination of size 50mm at 210mm from left
20	5	0	5	0	4.43366E-05	0.000028817	5.78142E-05	
21	0	5	5	0	4.22971E-05	0.000032214	4.04562E-05	
22	5	5	5	0	4.43376E-05	0.00003069	5.78142E-05	
23	0	5	0	0	8E-10	1.416E-06	1E-10	
24	5	5	0	0	8.44E-07	3.272E-06	1.35357E-05	
25	0	0	0	5	2.61841E-05	0.000035177	1.21353E-05	
26	5	0	0	5	0.000027744	3.2725E-05	5.221E-07	
27	0	0	5	5	2.61841E-05	0.000035177	1.21353E-05	
28	5	0	5	5	2.09678E-05	0.000103816	3.75002E-05	
29	0	5	5	5	2.13567E-05	0.000114459	2.26242E-05	
30	5	5	5	5	2.09723E-05	0.000107997	3.75012E-05	
31	0	5	0	5	2.61794E-05	0.000038553	1.21366E-05	
32	5	5	0	5	0.000027738	0.000036197	5.233E-07	

Table 4. Results of the DOE runs for the point-to-point analysis (A: Overlaminate – B: Bulkhead – C: Hull– D: Loading angle)

Factors	@first node		@middle node		@last node	
	Healthy	Damaged	Healthy	Damaged	Healthy	Damaged
A	0.13868	0.017835	0.043177	1.01244	4.91255	6.414703
B	8.15E-12	0.03857	0.117004	2.8447	1.95E-08	0.113719
C	**38.60064**	**38.8124**	**40.42626**	**33.93042**	**73.58083**	**66.59797**
D	**0.457217**	**0.827449**	**51.83837**	**42.73814**	**6.411934**	**8.096834**
AB	8.79E-11	0.038593	5.28E-05	1.694425	2.63E-10	0.113632
AC	0.000112	0.066135	0.054639	1.098861	3.868746	5.223829
AD	0.032741	0.112498	0.083222	0.983084	3.214382	2.070652
BC	2.67E-07	0.038292	0.000938	1.810764	9.47E-11	0.113615
BD	3.63E-08	0.038687	0.01842	2.126322	2.33E-08	0.113723
CD	**60.72826**	**59.65124**	**7.417231**	**3.218366**	**5.848022**	**7.467763**
ABC	2.51E-09	0.038622	6.35E-07	1.713334	2.63E-10	0.113615
ABD	1.15E-09	0.038568	1.08E-07	1.716435	1.05E-11	0.113619
ACD	0.04235	0.204216	0.000548	1.6494	2.163531	3.219085
BCD	2.87E-07	0.038282	0.000141	1.748334	5.16E-10	0.113611
ABCD	3.12E-09	0.038622	4.12E-08	1.714975	1.05E-11	0.113628

Table 5. Percentage contributions of the factors from the point-to-point analysis results in Table 4; all values are in %; the bold numbers refer to the high contributions.

Source	DF	Seq SS	Adj SS	Adj MS	F	P
A	1	291.69	291.69	291.69	20.44	0.002
C	1	4368.92	4368.92	4368.92	306.09	0.000
D	1	380.71	380.71	380.71	26.67	0.001
A*C	1	229.71	229.71	229.71	16.09	0.003
A*D	1	190.86	190.86	190.86	13.37	0.005
C*D	1	347.23	347.23	347.23	24.33	0.001
Error	9	128.46	128.46	14.27		
Total	15	5937	.57			

Table 6. Results of ANOVA for the 3rd node response, considering the identified factors from Figure 8 for the block of healthy runs

Next, based on the identified significant factors from the above results for the 3rd node, an ANOVA analysis (Table 6) was performed considering the rest of insignificant effects embedded in the error term. As expected, the p-value for the factor C is zero and the corresponding values for factors D and CD are 0.001. The p-value for all other factors is greater than 0.001. Therefore, assuming a significance level of 1%, for the 3rd node response, much like the 1st and middle nodes, factors C, D and their interaction CD can be reliably considered as most significant. Table 7 shows the ANOVA results for all the three nodes when only these three factors were included.

One interesting observation during the above analysis was that we found no significant deviation of main results when we repeated the analysis for the block of runs with delamination (compare the corresponding values under each node in Table 5 for the two healthy and damage cases). This indicated that *the effects of misalignment (manufacturing and testing error) factors between the healthy and damaged structures at each specific node are generally identical*, in the present case study.

Source	DF	Seq SS	Adj SS	Adj MS	F	P
@1ˢᵗ node						
C	1	1451.4	1451.4	1451.4	2165.69	**0.000**
D	1	17.2	17.2	17.2	25.65	**0.000**
C*D	1	2283.4	2283.4	2283.4	3407.16	**0.000**
Error	12	8.0	8.0	0.7		
Total	15	3759.9				
@Middle node						
C	1	10356.0	10356.0	10356.0	1524.83	**0.000**
D	1	13279.4	13279.4	13279.4	1955.28	**0.000**
C*D	1	1900.1	1900.1	1900.1	279.77	**0.000**
Error	12	81.5	81.5	6.8		
Total	15	25616.9				
@Last node						
C	1	4368.9	4368.9	4368.9	62.36	**0.000**
D	1	380.7	380.7	380.7	5.43	**0.038**
C*D	1	347.2	347.2	347.2	4.96	**0.046**
Error	12	840.7	840.7	70.1		
Total	15	5937.6				

Table 7. Results of ANOVA analysis for factors C, D and CD – point-to-point analysis approach

Figures 9.a – 96.f represent the main factor and interaction plots for the point-to-point analysis. For the first and last points, the lines for interaction of hull fiber misalignment and the loading angle offset are crossed, which indicates a high interaction between those parameters at the corresponding node. This interaction indication agrees well with the high F-value provided by the ANOVA analysis for CD in Table 7 for the first node. For the middle node, the individual lines for C and D in the main plots are in the same direction but with a small difference in their slopes. For the last (3ʳᵈ) node, the main factor plots for parameters C and D have slopes with opposing signs, suggesting that for this node, the fiber misalignment angle and loading angle offset have opposite influences on the strain response. This again could be explained by the imposed type of constraint on the right side of the T-joint.

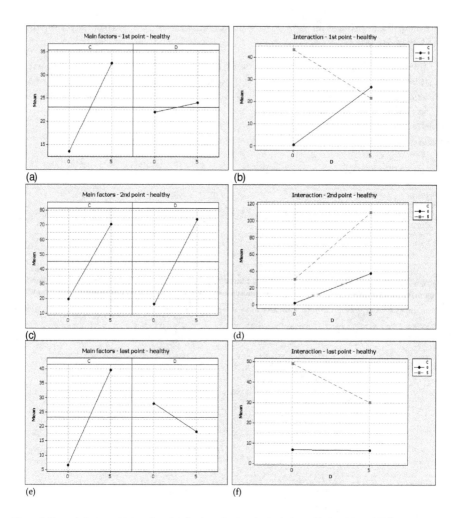

Figure 9. The main factor and interaction plots for the point-to-point analysis considering C, D and CD factors.

4.2. Integral analysis results

In this approach the objective function for each run was considered as the area between the curve representing the strain distribution of the nodes lying on the bond line and that of the base case. For the first group of runs (healthy structure, run#1-16), the first run is the base curve, whereas for the second group (embedded delamination case, run # 17 – 32) the 17th run (i.e., only delamination and no other fiber misalignment or loading angle error) is considered as the base. Table 8 lists the objective values for each run during this analysis. Table 9 represents the obtained percentage contribution of each factor. The parameters C and CD again play the main role on the strain distribution, but to be more accurate one may also consider other factors such as A, D, AD, and AC.

Run	A	B	C	D	Integral method response	Structure's state
1	0	0	0	0	0	
2	5	0	0	0	1.79422E-05	
3	0	0	5	0	4.33694E-05	
4	5	0	5	0	5.6027E-05	
5	0	5	5	0	4.33843E-05	
6	5	5	5	0	5.60299E-05	
7	0	5	0	5	3.73717E-05	No Delamination
8	5	5	0	5	3.14578E-05	
9	0	0	0	5	3.72118E-05	
10	5	0	0	5	3.14519E-05	
11	0	0	5	5	3.0095E-05	
12	5	0	5	5	3.77188E-05	
13	0	5	5	5	3.04696E-05	
14	5	5	5	5	3.80525E-05	
15	0	5	0	0	1.31538E-07	
16	5	5	0	0	1.79444E-05	
17	0	0	0	0	0	
18	5	0	0	0	1.79696E-05	
19	0	0	5	0	4.35521E-05	Delamination of size 50mm at 210mm from left
20	5	0	5	0	5.66002E-05	
21	0	5	5	0	4.35663E-05	
22	5	5	5	0	5.66009E-05	
23	0	5	0	0	1.32937E-07	
24	5	5	0	0	1.79724E-05	
25	0	0	0	5	3.7407E-05	
26	5	0	0	5	3.15725E-05	
27	0	0	5	5	3.7407E-05	
28	5	0	5	5	3.9519E-05	
29	0	5	5	5	3.28869E-05	
30	5	5	5	5	3.98452E-05	
31	0	5	0	5	3.7563E-05	
32	5	5	0	5	3.15788E-05	

Table 8. Results of the DOE runs for the integral analysis (A: Overlaminate – B: Bulkhead – C: Hull– D: Loading angle)

In order to show the dominant factors graphically, the corresponding half normal probability plot (Figure 10) was constructed; Figure 10 recommends considering AC as the last dominant factor. Next, a standard ANOVA analysis was performed (Table 10) and results suggested ignoring the effect of factors AC and D with a statistical significance level of $\alpha=0.01$. Nevertheless, recalling the percentage contributions in Table 9 it is clear that the top two main factors are C and CD, as it was the case for the point-to-point analysis. However in the point-to-point analysis, D was also highly significant at the selected nodes, whereas in the integral method it shows much less overall contribution. This would mean that *the number and locations of sensors during SHM can vary the sensitivity of the prediction results to particular noise/uncertainty factors,* such as D (the loading angle offset). Figure 11 illustrates the main and interaction plots for the factors A, C, and D. From Figure 11.a, unlike in the point-to-point analysis (Figure 9), the slope of every main factor, including D, is positive in the current analysis. This indicates that increasing each noise factor magnitude also increases the deviation of the structure's overall response from the base model. The interaction plot for C and D in Figure 11.b confirms an overall high interference of these two main factors; which is interesting because according to Figures 9 the lines of these factors cross each other mainly at the first node. This suggests that only for a few number of points near the left constraint point the interactive effect of noise factors (here C and D) may be notable; A potential hypothesis from these results for a future work would be: *the more dispersed the positions of the sensors, perhaps the less likelihood of imposing interactive effects of noise (uncertainty) factors on the overall prediction results.*

| Factors | Structure's health state | |
	Healthy	Damaged
A	**6.552184733**	**5.319084821**
B	0.001652565	0.02290527
C	**41.0309489**	**46.98622851**
D	**2.388791028**	**4.01506981**
AB	0.000177577	0.031520485
AC	0.423871764	0.189479205
AD	**5.197090974**	**6.353769585**
BC	0.000285795	0.030478626
BD	0.000820031	0.026597215
CD	**42.21458426**	**35.4607182**
ABC	8.33724E-05	0.039747913
ABD	4.48647E-06	0.035628966
ACD	2.188823941	1.425288598
BCD	0.000680545	0.027279407
ABCD	2.77746E-08	0.036203388

Table 9. Percentage contributions of the factors from the integral analysis in Table 8; all values are in %; the bold numbers refer to the high contributions

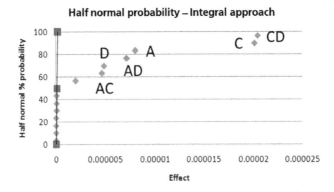

Figure 10. Half normal probability plot – integral approach (Healthy structure).

Source	DF	Seq SS	Adj SS	Adj MS	F	P
A	1	260.75	260.75	260.75	26.90	0.001
C	1	1632.87	1632.87	1632.87	168.43	0.000
D	1	95.06	95.06	95.06	9.81	0.012
A*C	1	16.87	16.87	16.87	1.74	0.220
A*D	1	206.82	206.82	206.82	21.33	0.001
C*D	1	1679.97	1679.97	1679.97	173.28	0.000
Error	9	87.25	87.25	9.69		
Total	15	3979.60				

Table 10. Results of ANOVA analysis based on dominant factors in Figure 10 for the integral approach

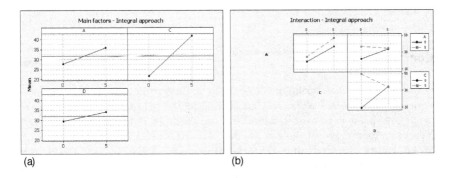

(a)   (b)

Figure 11. Main factor and interaction plots for the integral analysis approach (considering factors A, C, D and their interactions).

Finally, similar to the point-to-point analysis, comparing the contribution percentage values for the two blocks of runs in Table 9 (healthy vs. damaged structure), the delamination seems to have no major interaction with the other four uncertainty factors. In order to statistically prove this conclusion, a 2^5 full factorial was performed, considering delamination as a new fifth factor E. Ignoring the 3rd order interactions and embedding them inside the body of error term, ANOVA results were obtained in Table 11. It is clear that the E (damage) factor itself has a significant contribution but not any of its interaction terms with noise factors.

Source	DF	Seq SS	Adj SS	Adj MS	F	P
A	1	626.54	626.54	626.54	48.33	0.000
B	1	17.71	17.71	17.71	1.37	0.260
C	1	2817.34	2817.34	2817.34	217.30	0.000
D	1	274.12	274.12	274.12	21.14	0.000
E	**1**	**12.22**	**12.22**	**12.22**	**0.94**	**0.346**
A*B	1	12.82	12.82	12.82	0.99	0.335
A*C	1	28.51	28.51	28.51	2.20	0.158
A*D	1	1.49	1.49	1.49	0.12	0.739
A*E	1	13.53	13.53	13.53	1.04	0.322
B*C	1	13.22	13.22	13.22	1.02	0.328
B*D	1	16.58	16.58	16.58	1.28	0.275
B*E	1	13.06	13.06	13.06	1.01	0.331
C*D	1	1.02	1.02	1.02	0.08	0.782
C*E	1	12.65	12.65	12.65	0.98	0.338
D*E	1	13.08	13.08	13.08	1.01	0.330
Error	16	207.44	207.44	12.97		
Total	31	4081.34				

Table 11. Results of ANOVA analysis considering delamination as the 5th factor for integral approach.

5. Conclusions

Two different approaches, a point-to-point analysis and an integral analysis, were considered in a case study on the potential effect of uncertainty factors on SHM predictability in composite structures. The point–to-point (discrete) analysis is more similar to real applications where the number of sensors is normally limited and the SHM investigators can only rely on the data extracted at specific sensor locations. The integral approach, on the other hand, calculates the area of a continuous strain distribution and, hence, simulates an ideal situation where there are a very large number of sensors embedded inside the structure. The comparison of the two approaches showed the impact of increasing the number of strain measurement points on the behavior of the prediction model and the associated statistical results. Namely, for all sensor positions considered in the point-to-point (discrete) analysis, the main factors were the misalignment of fibers in the hull and the loading angle offset, but for the integral (continuous) approach, the aggregation of smaller factors over the bond line resulted in increasing significance of other parameters such as overlaminate misalignment angle and its interaction with other factors. However the top contributing factors remained the same between the two analyses, indicating that increasing the number of sensors does not eliminate the noise effects from fabrication such as misalignment of fibers and loading angle offset. Another conclusion from this case study was that, statistically, there was no sign of significant deviation in contribution patterns of factors between the healthy and damaged structure. This suggests that different sensor positioning scenarios may change the sensitivity of the response to noise factors but the deviation would be regardless of the absence or presence of delamination. In other words the relative importance of studied noise factors would be nearly identical in the healthy and damaged structure. Finally, results suggested that that the absolute effect of individual manufacturing uncertainty factors in deviating the structure's response can be as high as that caused by the presence of delamination itself when compared to the response of the healthy case, even in the absence of misalignment errors. Hence, a basic SHM damage prediction system under the presence of pre-existing manufacturing/testing errors may lead to wrong decisions or false alarms. A remedy to this problem is the use of new stochastic SHM tools.

Acknowledgements

Authors are grateful to Dr. J. Loeppky from Irving K. Barber School of Arts and Sciences at UBC Okanagan for his useful suggestions and discussions. Financial support from the Natural Sciences and Engineering Research Council (NSERC) of Canada is also greatly acknowledged.

Author details

H. Teimouri, A. S. Milani* and R. Seethaler

*Address all correspondence to: abbas.milani@ubc.ca

School of Engineering, University of British Columbia, Kelowna, Canada

References

[1] Christian BOLLER, Norbert MEYENDORF. "State-of-the-Art in Structural Health Monitoring for Aeronautics" Proc. of Internat. Symposium on NDT in Aerospace, Fürth/Bavaria, Germany, December 3-5, 2008.

[2] Charles R. FARRAR, Keith WORDEN. "An Introduction to Structural Health Monitoring" Phil. Trans. R. Soc. A 365, 303–315, 2007.

[3] C. Boller, M. Buderath. "Fatigue in aerostructures – where structural health monitoring can contribute to a complex subject" Phil. Trans. R. Soc. A 365, 561–587, 2007.

[4] R. Ikegami. "Structural Health Monitoring: assessment of aircraft customer needs" Structural Health Monitoring 2000, Proceedings of the Second International Workshop on Structural Health Monitoring, Stanford, CA, September 8-10, Lancaster – Basel, Technomic Publishing Co, Inc, pp. 12-23, 1999.

[5] J.M.W. BROWNJOHN. "Structural Health Monitoring of Civil Infrastructure" Phil. Trans. R. Soc. A 365, 589-622, 2007.

[6] Light Structures AS. "Exploration & Production: The Oil & Gas Review, Volume 2" 2003.

[7] Ved Prakash SHARMA, J.J. Roger CHENG. "Structural Health Monitoring of Syncrude's Aurora II Oil Sand Crusher" University of Alberta, Department of Civil & Enviromental Engineering, Structural Engineering Report No 272, 2007.

[8] D. BARKE, W.K. CHIU. "Structural Health Monitoring in Railway Industry: A Review" Structural Health Monitoring 4, 81-93, 2005.

[9] B.R. RANDALL. "State of the Art in Monitoring Rotating Machinery – Part I" J. Sound and Vibrations 38, March, 14-21, 2004.

[10] B.R. RANDALL. "State of the Art in Monitoring Rotating Machinery – Part II" J. Sound and Vibrations 38, May, 10-17, 2004.

[11] B. Tang, "Orthogonal array-based Latin hypercubes" Journal of the American Statistical Association, 88(424):1392–1397, 1993.

[12] J.-S. Park, "Optimal Latin-hypercube designs for computer experiments" Journal of Statistical Planning and Inference, 39:95–111, 1994.

[13] X. Qu, G. Venter, R.T. Haftka, "New formulation of minimumbias central composite experimental design and Gauss quadrature" Structural and Multidisciplinary Optimization, 28:231–242, 2004.

[14] R. Jin, W. Chen, A. Sudjianto, "An efficient algorithm for constructing optimal design of computer experiments" Journal of Statistical Planning and Inference, 134:268–287, 2005.

[15] T.W. Simpson, D.K.J. Lin, W. Chen, "Sampling strategies for computer experiments: Design and analysis" International Journal of Reliability and Applications, 2(3):209–240, 2001.

[16] R.H. Myers, D.C. Montgomery, "Response Surface Methodology—Process and Product Optimization Using Designed Experiments" New York: Wiley, 1995.

[17] A. Kesavan, S. John, I. Herszberg, "Strain-based structural health monitoring of complex composite structures" Structural Health Monitoring, 7:203-213, 2008.

Chapter 3

Optimization of Multiple Responses in Quench Hardening and Tempering Process in Steel Wires by Means Design of Experiments and Desirability Method

Cristie Diego Pimenta, Messias Borges Silva, Rosinei Batista Ribeiro,
Roberto Campos Leoni, Ricardo Batista Penteado,
Fabrício Maciel Gomes and Valério Antonio Pamplona Salomon

Additional information is available at the end of the chapter

1. Introduction

This research aims to show factors influence study and optimization of multiple mechanical properties responses of thermal treatment process quench hardening and tempering in steel wires used in manufacturing automotive springs. For the data collection and process statistical modeling, it was used the following methodologies: design of experiments and multiple linear regression. In this case, these methods were used to assist in a statistical modeling development which might replace the traditional way to adjust the input variables of thermal treatment process. This process setup is currently done by means of mechanical tests of pilot samples which is referred to laboratory analysis, after going by all stages of a thermal treatment for quenching hardening and tempering. Results obtained in this stage, are used to regulate the annealing furnace, implying considerable analysis and standby time, reducing, this way, the process productivity.

2. Bibliografic review

2.1. Thermal treatment and mechanical tests

According to Mayers and Chawla (1982), in a tensile strength test, the specimen is fixed on a testing machine head, which applies an effort that tends to elongate it up to rupture, where deformations are measured by means of a device called extensometer. The test is carried out on a specimen with standardized dimensions, so that the obtained results can be compared, reproduced and quantified at the machine itself. Normally, the test occurs up to the material failure (which is classified as destructive) and allows measuring the material strength and deformation depending on the applied strain. Above a certain strain level, materials start to deform plastically until there is a rupture, point where it is obtained the traction resistance limit. Universal testing machine for traction is the most used and the most common force units are kilogram-force per square millimeter (Kgf/mm^2) or MegaPascal (MPa).

Yield is the attribute presented by certain materials when undergoing large plastic transformations before their break when subjected to traction tension. In steel specimens, yield is measured by reduction of cross-sectional area which occurs before rupture. Yield is given by the ratio between variation of cross-sectional area of specimen (initial area - final area) and the value of initial area of cross-section (MAYERS; Chawla, 1982). Yield or area reduction is usually expressed as a percentage, showing how much of cross-sectional area of resistive section of specimen was reduced after force application in tensile test.

According to Callister (2002), hardness is a metal resistance measure to penetration. The most common methods to determine a metal hardness are Brinell, Vickers and Rockwell. In this research, only the Brinell method (BH) is used. Brinell hardness values (BH), as shown in Figure 1, are calculated by dividing applied load by penetration area. The diameter penetrator (D) is a hardened steel ball for materials of medium or low hardness, or tungsten carbide for high hardness materials. The test machine has a light microscope which makes the circle diameter measurement (d, in mm), which corresponds to the spherical cap projection printed on the sample. Brinell hardness (BH) is given by the applied load (P, in kgf) divided by the print area, as shown in equation 1.

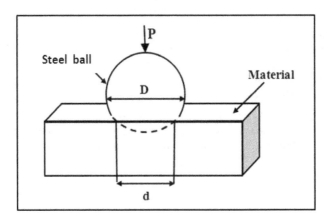

Figure 1. Brinell hardness (BH) method Illustration.

Source: Authors elaboration.

$$BH = \frac{2P}{\pi D \left(D - \sqrt{D^2 - d^2} \right)} \quad \left[\text{kgf}/\text{mm}^2 \right] \tag{1}$$

2.2. Statistical methods used

According to Lima et al. (2011), Silva and Silva (2008) and Granato et al. (2011), the design of experiments (DOE) is very adequate to study several process factors and their interactions complexity in order to solve problems by means of statistical analysis. According to Montgomery (2010) and Benyounis and Olabi (2008), blocking is a technique used to improve comparison accuracy among interest factors and can be used in conjunction with the multiple linear regression technique for process statistical modeling. Blocking can be employed in factorial planning when it is necessary to control variability coming from disturbing sources known, which may influence the results.

Montgomery and Runger (2003) state that multiple linear regression is used for situations involving more than one regressor, and the models can include interaction effects. An interaction between two variables can be represented by a cross term, for if we assume that x_3 = $x_1 x_2$ and $\beta_3 = \beta_{12}$, then the model, including interaction terms, will be as shown in equation 2.

$$Y = \beta_0 + \beta_1 x_1 + \beta_2 x_2 + \beta_3 x_3 + \ldots + \varepsilon \tag{2}$$

In this expression, Y is the dependent variable; the independent variables are represented by $x_1, x_2, \ldots, x_n$ and ε is the random error term. The term "linear" is used because the equation is a linear function of the unknown parameters $\beta_0, \beta_1, \beta_2$ and β_n. In this model, the parameter β_0 is the plane intersection; β_1, β_2 and β_n are the regression partial coefficients.

The desirability method is a method used for determining the best conditions for process adjustment, making possible simultaneous optimization of multiple responses. This being so, the best responses conditions are obtained simultaneously minimizing, maximizing or seeking nominal values of specifications, depending on the most convenient situation for the process (WANG, WAN, 2009).

Each one of responses ($Y_1, Y_2...Y_k$) of original set is transformed, such that d_i belongs to interval $0 \le d_i \le 1$. The d_i value increases when the ith response approaches the imposed limits. Equation 3 is used to find the D global index, from combination of each one responses processed through a geometric mean.

$$D = \left(d_1(Y_1) \times d_2(Y_2) \ldots \times d_k(Y_k) \right)^{\frac{1}{k}} \tag{3}$$

As a result of geometric mean represented by equation 3, the value D evaluates, in a general way, the levels of the combined set of responses. It is an index also belonging to interval [0, 1] and will be maximized when all responses approach as much as possible of its specifications. The closer of one D is, the closer the original responses will be of their respective specification limits. The general optimal point of system is the optimal point achieved by maximizing the geometric mean, calculated from individual desirability functions (Paiva, 2008). According to Paiva (2008), advantage of using geometric mean is to make the overall solution is achieved in a balanced way, allowing all responses can achieve the expected values and forcing algorithm to approach the imposed specifications.

According to Derringer and Suich (1980), the algorithm will depend on the optimization type desired for response (maximization, minimization or normalization) of desired limits within the specification and the amounts (weights) of each one response, which identifies the main characteristics of different optimization types, as follows:

- Minimize Function: The desirability function value increases as the original response value approaches a minimum target value;

- Normalize Function: When response moves toward the target, the desirability function value increases;

- Maximize Function: The desirability function value increases when the response value increases.

Paiva (2008) and WU (2005) state that when a response maximization is wished, the transformation formula is shown in equation 4:

$$
di = \begin{cases}
0 & \hat{Y}_i < LSL \\[2mm]
\left[\dfrac{\hat{Y}_i - Li}{Ti - Li} \right]^R & L_i \le \hat{Y}_i \le T_i \\[2mm]
1 & \hat{Y}_i > T_i
\end{cases} \tag{4}
$$

Where: L_i, T_i and H_i are, respectively, the values of major, minor and acceptable target for the ith response.

The R value, in Equation 4, indicates a preponderance of the superior limit (LSL). Values higher than unity should be used when the response (Yi) increases rapidly above L_i. Therefore, d_i increases slowly, while the response value is being maximized. Consequently, to maximize D, the ith response must be much larger than L_i. One can choose R <1, when it is critical to find values for the response below the fixed limits.

In cases where the objective is to reach a target value, the transformation formulation stops being unilateral and becomes bilateral. The bilateral formulation, represented by equation 5, occurs when the interest response has two restrictions: one maximum and the other one minimum.

$$d_i = \begin{cases} 0 & \hat{Y}_i < L_i \, ou \, \hat{Y}_i > H_i \\[2ex] \left[\dfrac{H_i - \hat{Y}_i}{Hi - Ti} \right]^R & T_i \leq \hat{Y}_i \leq H_i \\[2ex] \left[\dfrac{\hat{Y}_i - Li}{Ti - Li} \right]^R & L_i \leq \hat{Y}_i \leq T_i \end{cases} \tag{5}$$

3. Materials and methods

3.1. Material, factors selection and experimental organization

The material used was SAE 9254 steel wire, with diameter gauges 2.00 mm and 6.50 mm. Factors investigated in this research are:

- Speed of wire passage inside the furnace (in m/s);

- Polymer concentration, quenching medium (in %);

- Lead temperature in tempering (in °C).

The steel wire diameter was also considered as an important factor, for there was assumption that its mass could influence results of investigated mechanical properties. Nevertheless, in this research, it was used the blocks analysis methodology, that is, for block 1, it was allocated experiments related only to diameter 2.00 mm, and for block 2, experiments related to 6.50 mm diameter as shown in Table 1.

Experiments	Speed	Lead Temperature	% Polymer
1	-	-	-
2	+	-	-
3	-	+	-
4	+	+	-
5	-	-	+
6	+	-	+
7	-	+	+
8	+	+	+

Table 1. Factorial Matrix 2^3

Factors such as speed, lead temperature and polymer concentration were tested by means of the factorial planning, using the matrix 2^3.

For experiments planning accomplishment, reduced variables (β) were used rather than physical variables (real adjustments) of investigated factors, in order to preserve the confidential data of the company which funds the research. Variables reduction was calculated according to Montgomery and Runger (2003), using the physical value (α) that one wants to test subtracted from the mean (μ) between the minimum and maximum of factors adjustments. The result was divided by half the amplitude (R) between the minimum and maximum values of factors adjustment. Thus, the reduced variables dimensionality was restricted to the range [-1 to 1], according to equation 6 and Table 2.

$$\beta = \frac{\alpha - \mu}{\dfrac{R}{2}}$$ (6)

Input variables	Values (physical units)	Values (reduced variables)
Speed (m/s)	Minimum / Maximum	-1 / 1
Lead temperature (°C)	Minimum / Maximum	-1 / 1
Polymer concentration (%)	Minimum / Maximum	-1 / 1

Table 2. Transformation of physical variables to reduced variables

4. Results and discussion

4.1. Sequence of experiments and statistical analysis

In the experiments, all replicas related to block 1 were initially carried out, and then the ones corresponding to block 2. Six replicas were used for each experimental condition. Replications were randomized and sequenced using a notation from 1 to 8, corresponding to each experiment order for each block individually. This experimental sequence is displayed in parentheses and in subscript format next to values obtained from mechanical properties as displayed in Tables 3, 4 and 5.

Experiments	Replica 1	Replica 2	Replica 3	Replica 4	Replica 5	Replica 6
1/Block 1	2149 (1)	2148 (1)	2146 (2)	2161 (8)	2167 (1)	2160 (6)
2/Block 1	2157 (4)	2155 (7)	2157 (3)	2151 (7)	2157 (4)	2157 (2)
3/Block 1	1924 (3)	1922 (3)	1920 (1)	1921 (5)	1920 (6)	1918 (4)

Experiments	Replica 1	Replica 2	Replica 3	Replica 4	Replica 5	Replica 6
4/Block 1	$1924_{(2)}$	$1924_{(8)}$	$1922_{(8)}$	$1943_{(6)}$	$1945_{(8)}$	$1945_{(5)}$
5/Block 1	$2108_{(6)}$	$2106_{(5)}$	$2108_{(7)}$	$2104_{(2)}$	$2102_{(7)}$	$2109_{(8)}$
6/Block 1	$2136_{(5)}$	$2127_{(4)}$	$2127_{(4)}$	$2136_{(3)}$	$2134_{(3)}$	$2127_{(3)}$
7/Block 1	$1927_{(7)}$	$1926_{(2)}$	$1944_{(5)}$	$1935_{(4)}$	$1946_{(2)}$	$1947_{(7)}$
8/Block 1	$1946_{(8)}$	$1946_{(6)}$	$1946_{(6)}$	$1953_{(1)}$	$1951_{(5)}$	$1946_{(1)}$
1/Block 2	$1968_{(1)}$	$1974_{(1)}$	$1962_{(3)}$	$1971_{(4)}$	$1971_{(8)}$	$1974_{(5)}$
2/Block 2	$1980_{(7)}$	$1976_{(4)}$	$1988_{(6)}$	$1978_{(2)}$	$1980_{(3)}$	$1988_{(2)}$
3/Block 2	$1771_{(3)}$	$1764_{(3)}$	$1763_{(7)}$	$1773_{(5)}$	$1771_{(5)}$	$1764_{(4)}$
4/Block 2	$1796_{(8)}$	$1784_{(2)}$	$1797_{(8)}$	$1781_{(3)}$	$1796_{(2)}$	$1784_{(3)}$
5/Block 2	$1949_{(5)}$	$1963_{(6)}$	$1947_{(1)}$	$1951_{(1)}$	$1949_{(4)}$	$1947_{(6)}$
6/Block 2	$1992_{(4)}$	$1980_{(5)}$	$1976_{(4)}$	$1994_{(8)}$	$1980_{(7)}$	$1992_{(7)}$
7/Block 2	$1760_{(2)}$	$1768_{(7)}$	$1766_{(5)}$	$1763_{(7)}$	$1766_{(6)}$	$1763_{(8)}$
8/Block 2	$1787_{(6)}$	$1793_{(8)}$	$1785_{(2)}$	$1784_{(6)}$	$1784_{(1)}$	$1785_{(1)}$

Table 3. Tensile strength results (MPa)

Experiments	Replica 1	Replica 2	Replica 3	Replica 4	Replica 5	Replica 6
1/Block 1	$50_{(1)}$	$51_{(1)}$	$51_{(2)}$	$50_{(8)}$	$50_{(1)}$	$50_{(6)}$
2/Block 1	$50_{(4)}$	$50_{(7)}$	$50_{(3)}$	$50_{(7)}$	$50_{(4)}$	$50_{(2)}$
3/Block 1	$58_{(3)}$	$58_{(3)}$	$58_{(1)}$	$58_{(5)}$	$58_{(6)}$	$58_{(4)}$
4/Block 1	$58_{(2)}$	$58_{(8)}$	$58_{(8)}$	$56_{(6)}$	$56_{(8)}$	$56_{(5)}$
5/Block 1	$53_{(6)}$	$53_{(5)}$	$53_{(7)}$	$53_{(2)}$	$53_{(7)}$	$53_{(8)}$
6/Block 1	$51_{(5)}$	$52_{(4)}$	$52_{(4)}$	$51_{(3)}$	$51_{(3)}$	$52_{(3)}$
7/Block 1	$58_{(7)}$	$58_{(2)}$	$56_{(5)}$	$58_{(4)}$	$56_{(2)}$	$56_{(7)}$
8/Block 1	$56_{(8)}$	$56_{(6)}$	$56_{(6)}$	$55_{(1)}$	$56_{(5)}$	$56_{(1)}$
1/Block 2	$42_{(1)}$	$41_{(1)}$	$42_{(3)}$	$42_{(4)}$	$42_{(8)}$	$41_{(5)}$
2/Block 2	$41_{(7)}$	$41_{(4)}$	$40_{(6)}$	$41_{(2)}$	$41_{(3)}$	$40_{(2)}$
3/Block 2	$47_{(3)}$	$46_{(3)}$	$46_{(7)}$	$47_{(5)}$	$47_{(5)}$	$46_{(4)}$
4/Block 2	$44_{(8)}$	$45_{(2)}$	$44_{(8)}$	$45_{(3)}$	$44_{(2)}$	$45_{(3)}$
5/Block 2	$56_{(5)}$	$42_{(6)}$	$56_{(1)}$	$56_{(1)}$	$56_{(4)}$	$56_{(6)}$
6/Block 2	$40_{(4)}$	$41_{(5)}$	$41_{(4)}$	$40_{(8)}$	$41_{(7)}$	$40_{(7)}$
7/Block 2	$46_{(2)}$	$47_{(7)}$	$47_{(5)}$	$46_{(7)}$	$47_{(6)}$	$46_{(8)}$
8/Block 2	$44_{(6)}$	$44_{(8)}$	$45_{(2)}$	$45_{(6)}$	$45_{(1)}$	$45_{(1)}$

Table 4. Yield point results in percentage (%)

Experiments	Replica 1	Replica 2	Replica 3	Replica 4	Replica 5	Replica 6
1 / Block 1	$608_{(1)}$	$606_{(1)}$	$606_{(2)}$	$611_{(8)}$	$611_{(1)}$	$611_{(6)}$
2 / Block 1	$608_{(4)}$	$608_{(7)}$	$608_{(3)}$	$608_{(7)}$	$608_{(4)}$	$608_{(2)}$
3 / Block 1	$544_{(3)}$	$542_{(3)}$	$542_{(1)}$	$542_{(5)}$	$542_{(6)}$	$542_{(4)}$
4 / Block 1	$544_{(2)}$	$544_{(8)}$	$542_{(8)}$	$550_{(6)}$	$550_{(8)}$	$550_{(5)}$
5 / Block 1	$594_{(6)}$	$594_{(5)}$	$594_{(7)}$	$594_{(2)}$	$594_{(7)}$	$594_{(8)}$
6 / Block 1	$603_{(5)}$	$600_{(4)}$	$600_{(4)}$	$603_{(3)}$	$603_{(3)}$	$600_{(3)}$
7 / Block 1	$544_{(7)}$	$544_{(2)}$	$550_{(5)}$	$547_{(4)}$	$550_{(2)}$	$550_{(7)}$
8 / Block 1	$550_{(8)}$	$550_{(6)}$	$550_{(6)}$	$553_{(1)}$	$550_{(5)}$	$550_{(1)}$
1 / Block 2	$556_{(1)}$	$558_{(1)}$	$556_{(3)}$	$556_{(4)}$	$556_{(8)}$	$558_{(5)}$
2 / Block 2	$558_{(7)}$	$558_{(4)}$	$561_{(6)}$	$558_{(2)}$	$558_{(3)}$	$561_{(2)}$
3 / Block 2	$500_{(3)}$	$497_{(3)}$	$497_{(7)}$	$500_{(5)}$	$500_{(5)}$	$497_{(4)}$
4 / Block 2	$508_{(8)}$	$503_{(2)}$	$508_{(8)}$	$503_{(3)}$	$508_{(2)}$	$503_{(3)}$
5 / Block 2	$550_{(5)}$	$556_{(6)}$	$550_{(1)}$	$550_{(1)}$	$550_{(4)}$	$550_{(6)}$
6 / Block 2	$564_{(4)}$	$558_{(5)}$	$558_{(4)}$	$564_{(8)}$	$558_{(7)}$	$564_{(7)}$
7 / Block 2	$497_{(2)}$	$500_{(7)}$	$500_{(5)}$	$497_{(7)}$	$500_{(6)}$	$497_{(8)}$
8 / Block 2	$506_{(6)}$	$506_{(8)}$	$503_{(2)}$	$503_{(6)}$	$503_{(1)}$	$503_{(1)}$

Table 5. Hardness results (Brinell Hardness)

Factors significance was tested at a 95% confidence level ($p < 0.05$). This analysis was carried out separately so that factors significance for each response of studied mechanical properties could be verified, as shown in Tables 6, 7 and 8.

Terms	Effect	Coefficient	T	p
Constant		1955.29	1782.89	0.000
(D)	165.62	82.81	80.09	0.000
(A)	17.42	8.71	7.94	0.000
(B)	-198.54	-99.27	-90.52	0.000
(C)	-8.04	-4.02	-3.67	0.000
(A)(B)	-0.54	-0.27	-0.25	0.805
(A)(C)	5.62	2.81	2.56	0.012
(B)(C)	14.08	7.04	6.42	0.000
(A)(B)(C)	-6.25	-3.13	-2.85	0.005

Table 6. Significance test for resistance limit, by means of the Minitab Statistical Software (in MPa)

By means of the significance test performed for the mechanical property called tensile strength (shown in Table 6), it was found that the significant factors (where $p < 0.05$) are: wire diameter (represented by letter D and tested by means of Blocks), speed (represented by letter A), lead temperature (represented by letter B), polymer concentration (represented by letter C), second order interactions among speed and polymer concentration, polymer concentration and temperature and a third-order interaction among speed, lead temperature and polymer concentration.

Terms	Effect	Coefficient	T	p
Constant		49.458	201.94	0.000
(D)	9.426	4.713	201.94	0.000
(A)	-2.750	-1.375	-5.61	0.000
(B)	3.583	1.792	7.32	0.000
(C)	1.750	0.875	3.57	0.001
(A)(B)	1.250	0.625	2.55	0.012
(A)(C)	-1.667	-0.833	-3.40	0.001
(B)(C)	-2.250	1.125	-4.59	0.000
(A)(B)(C)	1.667	0.833	3.40	0.001

Table 7. Significance test for yield, by means of the Minitab Statistical Software (in percentage)

When analyzing the significance test for the mechanical property Yield (shown in Table 7), it is possible to note that the influential factors (where $p < 0.05$) are: wire diameter (tested by blocks), speed, lead temperature, polymer concentration, second order interactions among speed and lead temperature, speed and polymer concentration, temperature and polymer concentration and a third-order interaction among speed, lead temperature and polymer concentration.

Terms	Effect	Coefficient	T	p
Constant		552.09	1650.05	0.000
(D)	46.86	23.43	74.26	0.000
(A)	4.85	2.43	7.25	0.000
(B)	-55.81	-27.91	-83.40	0.000
(C)	-2.19	-1.09	-3.27	0.001
(A)(B)	0.10	0.05	0.16	0.877
(A)(C)	1.65	0.82	2.46	0.016
(B)(C)	4.06	2.03	6.07	0.000
(A)(B)(C)	-2.35	-1.18	-3.52	0.001

Table 8. Significance test for hardness, by means of the Minitab Statistical Software (in BH)

Analyzing the significance test for hardness mechanical property (displayed in Table 8), it is possible to state that the influential factors (in which p <0.05) are: wire diameter (tested by means of blocks), speed, lead temperature, polymer concentration, second order interactions among speed and polymer concentration, temperature and polymer concentration and a third-order interaction between lead temperature and polymer concentration.

4.2. Statistical modeling for multiple responses

Using coefficients calculated using the significance test, by means of the Minitab Statistical Software, it was possible to build statistical models which represent the relationship between process input variables (factors) and output variables (mechanical properties). Such statistical models are defined in equations 7, 8 and 9.

$$RL = 1955.29 + 82.81(D) + 8.71(A) - 99.27(B) - 4.02(C) + 2.81(A)(C) + 7.04(B)(C) - 3.13(A)(B)(C) \tag{7}$$

$$Y = 49.458 + 4.713(D) - 1.375(A) + 1.792(B) + 0.875(C) + 0.625(A)(B) - 0.833(A)(C) - 1.125(B)(C) + 0.833(A)(B)(C) \tag{8}$$

$$H = 552.09 + 23.43(D) + 2.43(A) - 27.91(B) - 1.09(C) + 0.82(A)(C) + 2.03(B)(C) - 1.18(A)(B)(C) \tag{9}$$

Where:

- RL: corresponds to the response variable called tensile strength;
- Y: corresponds to the variable called yield response;
- H: corresponds to the response variable called Hardness.

4.3. Application of *desirability* function for optimization

For process optimization by means of desirability function, firstly, it was necessary to formulate the specifications required for the studied mechanical properties. To this, blocks were analyzed separately, that is, the response variables were optimized primarily for the wire diameter 2.00 mm and then the same procedure was carried out to diameter 6.5 mm.

Specifications (minimum, nominal and maximum) concerning the diameter the 2.00 mm diameter are presented in Table 9. In that case, one seeks nominal values (target) for mechanical properties such as traction resistance limit and hardness and, for the mechanical property called yield, one seeks to maximization, for the higher the value, the better the product itself.

The composite desirability (D) is the overall index calculated from combination of each response variables processed through a geometric mean and this index is responsible for showing the best condition to optimize all responses variables at the same time. To obtain the highest possible value for D, which reflects in the best condition of response variables in relation to their specifications care (displayed in Figure 2), the best adjustments using factors reduced variables [-1 to 1] are:

Tensile strength (MPa)			Yield (%)			Hardness (BH)		
Minimum	Nominal (target)	Maximum	Minimum	Nominal	Maximum (target)	Minimum	Nominal (target)	Maximum
1930	2040	2150	40	45	≥ 50	545	572	600

Table 9. Specifications for 2.00 mm gauge

- Speed, fit in -1.0;

- Lead temperature fit in -0.0909;

- Polymer concentration fit in 1.0.

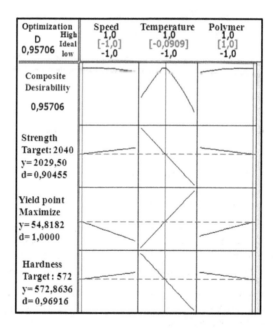

Figure 2. Desirability function applied in multiple responses (Minitab Statistical Software-2.00 mm diameter)

Looking at Figure 2, it can be seen that D value belonging to [0-1] interval, is maximized when all responses are close to their specifications, for the closer D is of 1, the closer the original responses will be of their respective specification limits. The optimal general point of the system is the optimum point achieved by geometric mean maximization calculated from individual desirability functions (d), which in this case are values for each one of response variables given below:

- For response variable called tensile strength, d=0.90455;

- For response variable called yield, d=1.0;

- For response variable called hardness, d=0.96916.

Values obtained for desirability (D) and individual desirability (d), show that the process was well optimized, since these indices are found to be very close to the optimum condition (1.0). Thus, it was possible to find that values obtained for this optimized condition are in accordance with required specifications and are:

- For tensile strength (y= 2029.5 MPa);

- For yield (y= 54.8182 %);

- For hardness (y= 572.8636 BH).

By analyzing Figure 2, it was found that speed factor, when increased, also causes increased amounts of response variables tensile strength (MPa) and hardness (BH). Also, the increased speed affects yield response variable reduction (%) and desirability (D) composite reduction.

Regarding the lead temperature factor, with increasing temperature, one realizes values reduction of response variables tensile strength (MPa), Hardness (BH) and desirability composite (D). On the other hand, yield value increases (%).

By observing increase in polymer concentration factor, one can see that there will be decrease in response variables values called tensile strength (MPa) and hardness (BH), yield increase (%) and desirability composite (D).

In Table 10, it is shown specifications (minimum, nominal and maximum) relative to 6.50 mm diameter. Also one searches nominal values (target) for mechanical properties called tensile strength and hardness, and for mechanical property called yield, one seeks maximization.

Traction resistance limit (MPa)			Yield (%)			Hardness (BH)		
Minimum	Nominal (target)	Maximum	Minimum	Nominal	Maximum (target)	Minimum	Nominal (target)	Maximum
1770	1875	1980	40	48	≥ 56	500	530	560

Table 10. Specifications for 6.50 mm gauge

As shown in Figure 3, for obtaining the highest possible value for *desirability* composite (D), the best factors adjustments are:

- Speed, fit at -1.0;

- Lead temperature, fit at -0.1919;

- Polymer concentration, fit at 1.0.

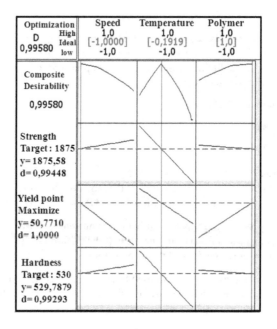

Optimization	Speed	Temperature	Polymer
High	1,0	1,0	1,0
D Ideal	[-1,0000]	[-0,1919]	[1,0]
0,99580 low	-1,0	-1,0	-1,0
Composite Desirability 0,99580			
Strength Target : 1875 y= 1875,58 d= 0,99448			
Yield point Maximize y= 50,7710 d= 1,0000			
Hardness Target : 530 y= 529,7879 d= 0,99293			

Figure 3. *Desirability* function applied in multiple responses (Minitab Statistical Software- 6.50 mm diameter)

By Figure 3 analysis, it is possible to realize that:

- For response variable called tensile strength, d=0.99448;

- For response variable called yield, d=1.0;

- For response variable called Hardness, d=0.99293.

It is also possible to observe that values obtained for this optimized condition comply with required specifications, which are:

- For tensile strength, (y= 1875.5791 MPa);

- For yield, (y= 50.7710 %);

- For hardness, (y= 529.7879 BH);

Regarding the speed factor, by increasing the speed one obtains values increase of response variable called tensile strength (MPa) and hardness (BH). Also, with increasing speed factor, it is observed a response variable reduction called yield (%) and desirability composite reduction (D).

Regarding the lead temperature factor, the increase means that there is all response variables decrease, including the desirability composite (D). By observing the polymer concentration factor, it is found that the increase will cause decrease of response variables tensile strength and hardness, increasing yield and desirability composite (D).

The red line (vertical) contained in Figure 3, can be interpreted as follows: in case it is moved, it will change the response values, and this will directly affect the composite desirability (D) values and individual desirability (d). For instance, by moving the red line, contained in the space relative to the lead temperature factor to the right, it will provide drop in the desirability composite (D), and all response variables (shown in Figure 3). It is possible to realize the drop in desirability composite (D) by observing the slope of straight contained in the location indicated previously. This decrease in D would represent optimization reduction of multiple responses and consequently no use of responses at their best factors adjustment conditions.

5. Conclusions

The design of experiments methodology with analysis in blocks applied to quench hardening and tempering process in SAE 9254 drawn steel wires with 2.00 mm and 6.50 mm diameters provided a wide understanding of factors influence in mechanical properties called tensile strength, hardness and yield.

By means of significance test (through of the Minitab Statistical Software), it was possible to find that factors such as diameter, speed, tempering temperature and polymer concentration have a significant influence on the studied mechanical properties and by statistical methods application it was possible to model the process, obtaining the best factors adjustment condition, which in turn, provided simultaneously multiple responses optimization.

Through the findings generated by this study, one seeks to fit in a planned way the quench hardening furnace set-up in a productive environment, obtaining, this way, reduction of initial laboratory tests amount and waiting time of these results, whose cost impacts directly the company financial indicators.

Author details

Cristie Diego Pimenta[1], Messias Borges Silva[1], Rosinei Batista Ribeiro[2], Roberto Campos Leoni[1], Ricardo Batista Penteado[1], Fabrício Maciel Gomes[1] and Valério Antonio Pamplona Salomon[1]

*Address all correspondence to: pimentadiego@yahoo.com

1 FEG UNESP – SP / Brasil, Guaratinguetá-SP, Brasil

2 FATEA – SP / Brasil, Lorena – SP, Brasil

References

[1] Benyounis, K. Y, & Olabi, A. G. Optimization of different welding processes using statistical and numerical approaches- A reference guide, *Science Direct*, (2008). , 39, 483-496.

[2] Callister, J. R. W. D. *Uma introdução a engenharia e a ciências dos materiais*, 5ª edição, editora LTC, (2002). , 589.

[3] Derringer, G, & Suich, R. Simultaneous Optimization of Several Response Variables, Journal of Quality Technology, n 4, (1980). , 12, 214-219.

[4] Granato, D, Branco, G. F, & Calado, V. M. A. Experimental design and application of response surface methodology for process modeling and optimization: A review, *Food Research International*, (2011). , 1, 0-14.

[5] Lima, V. B. S, Balestrassi, P. P, & Paiva, A. P. Otimização do desempenho de amplificadores de radio frequência banda larga: uma abordagem experimental, *Produção*, n. 1, jan/mar, (2011). , 21, 118-131.

[6] Mayers, A. M, & Chawla, K. K. *Princípios de metalurgia mecânica*, 2ª edição, Edgard Blucher, (1982). p.

[7] Montgomery, C. D. *Design and analysis of experiments*, 7ᵗʰ edition, John Wiley & Sons, (2010). p.

[8] Montgomery, D. C, & Runger, G. C. *Estatística aplicada e probabilidade para engenheiros*, 2ª edição, editora LTC, (2003). p., 2003, 230-320.

[9] Paiva, E. J. *Otimização de Manufatura com Múltiplas Respostas baseadas em índices de capacidade*, Dissertação, Universidade Federal de Itajubá, (2008). p.

[10] Silva, H. A, & Silva, M. B. Aplicação de um projeto de Experiments (DOE) na soldagem de tubos de zircaloy-4; *Produção & Engenharia*, n. 1, set./dez. (2008). , 1, 41-52.

[11] Wang, J, & Wan, W. Application of desirability function based on neural network for optimizing biohydrogen production process, *international journal o f hydrogen energy*, (2009). , 34, 1253-1259.

[12] Wu, F. C. optimization of correlated multiple quality characteristics using desirability function. *Quality engineering*, n 1, (2005). , 17, 119-126.

Chapter 4

Influential Parameters to the Database Performance — A Study by Means of Design of Experiments (DoE)

Eduardo Batista de Moraes Barbosa and
Messias Borges Silva

Additional information is available at the end of the chapter

1. Introduction

The growth of the Organization's data collections, due to the development of new materials and advanced computing devices, puts the database (Db) technology at the forefront. Consequentially to its popularity, different options of database management systems (DBMS) can be found on the market to store one of the most important Organization's assets, the data. Among the factors that may influence its choice (advanced features, interoperability, etc.), could be highlighted the cost-benefit provided by fierce competition between different software philosophies – proprietary (Oracle, MS SQL Server, IBM DB2, etc.) and public domain (PostgreSQL, MySQL, Firebird, etc.) – and, also the system performance in critical computing environments.

Performance measurements in the computer systems area (processors, operating systems, compilers, database systems, etc.) are conducted through benchmarks (a standardized problem or test used as basis for evaluation or comparison) widely recognized by the industry. There are a lot of the benchmarks consortia with specific evaluation criteria, metrics, pricing and results communication, highlighting: Open Source Database Benchmark (OSDB) (http://osdb. sourceforge.net), System Performance Evaluation Cooperative (SPEC) (http://www.spec.org) and Transaction Processing Performance Council (TPC) (http://www.tpc.org).

In the academic scope, the TPC benchmarks are widely recognized [1; 2; 3; 4; 5; 6; 7] due its exactness for definition of tests implementations, price measurements and results reporting. The TPC began from two *ad hoc* benchmarks formalization (DebitCredit and TP1) which resulted in the TPC BM™ and the TPC BM™ B [1]. Currently, with the benchmarks advance, its possible performs complex queries, batch and operational aspects of systems for transactions processing through different benchmark standards, such as TPC-C (simulates a complete computing environment where a population of users executes transactions against a Db), TPC-DS (models several generally applicable aspects of a decision support system), TPC-E (uses a database to model a brokerage firm with customers who generate transactions related to the business), TPC-H (consists of a suite of business oriented *ad-hoc* queries and concurrent data modifications), TPC-VMS (leverages the TPC benchmarks by adding the methodology and requirements for running and reporting performance metrics for virtualized databases) and TPC-Energy (contains the rules and methodology for measuring and reporting an energy metric in TPC benchmarks).

In this chapter we intend to present a study on the Db performance by means of statistical techniques for planning and analysis of experiments (DoE), applied in the computing scope. The objective of this study is use two DoE techniques (2^k full factorial and 2^{k-p} fractional factorial) to investigates the influence of different parameters (related with Db memory tuning) in the Db performance. The DoE methodology will be applied at the case study, where the Db parameters will be simultaneously combined and tested and its results analyzed by means of full factorial and fractional factorial designs, and to assist in the investigations to determine how each parameter may explain (or take influence in) the Db performance. Thus, will be also addressed a comparison of results between the both techniques chosen. It should be noted that, in the scope of this study, the Db technology will be used as a vehicle to demonstrate how the DoE methodology can help in the design of experiments and its analysis and its use as a promising tool in several scopes, like in the computing science field.

The paper is structured as follows: Section 2 shows an introduction in the benchmark technology, with emphasis on the TPC-H standard. The DoE methodology is introduced in Section 3, where can be found a overview over full factorial and fractional factorial designs. The Section 4 is devoted to the case study to investigate the influence of different Db parameters (from PostgreSQL DBMS) in its performance through DoE designs (2^k full factorial and 2^{k-p} fractional factorial). This Section also presents the analysis and comparison of results. Some related work are presented in Section 5 and the final considerations are in Section 6.

2. Benchmark overview

Performance tests in the computing scope are a valuable tool to assist the decision makers in the hardware and/or software settings. Such tests, usually called as benchmark, can ensure that software does not present problems or unavailability due to insufficient resources (i.e.: memory, processor, disk, etc.).

According to the Merriam-Webster dictionary (http://www.merriam-webster.com), benchmark is *"a standardized problem or test that serves as a basis for evaluation or comparison (as of computer system performance)"*. In the computing scope, benchmark is typically a software to perform pre-defined operations and returns a metric (i.e.: workload, throughput, etc.) to describe the system behavior.

Some generic benchmarks have become widely recognized (i.e.: TPC), such that vendors advertise new products performance based in its results. Benchmarks are important tools in evaluating computer system performance and price/performance. However, to be really useful [1] prescribes some key criteria that should be considered during the benchmarks' choice and/ or use:

- Relevance (must measure the peak performance and price/performance of systems when performing typical operations within that problem domain);

- Portability (should be easy to implement on many different systems and architectures);

- Scalability (should apply to small and large computer systems); and

- Simplicity (must be understandable).

Benchmark softwares, usually, simulates scenarios from the real world (i.e.: TPC-C, TPC-DS, TPC-E, TPC-H and TPC-VMS) where systematic procedures are performed to test, collect and analysis the system performance. It's use is strongly recommended, because the cost of implementing and measuring specific applications on different systems/platforms is usually prohibitive [1].

2.1. TPC-H Standard

TPC-H benchmark represent a generic decision support benchmark, with the main features: capacity to manage very large amounts of data, the power to analyze it with a high degree of complexity and the flexibility to answers critical business questions.

Figure 1, illustrates the logical schema of the TPC-H specification and shows the business and application environment. According to the TPC official documentation [8], *"TPC-H does not represent the activity of any particular business segment, but rather any industry which must manage, sell, or distribute a product worldwide (i.e.: car rental, food distribution, parts, suppliers, etc.)"*.

In this schema (Figure 1), the Db consists of eight tables simulating a realistic application involving customers, parts, lineitems, suppliers and orders. The prefix of the table columns is expressed into parentheses, the relationships between tables are represented by arrows and the number/formula below each table name are the cardinality (number of rows) of the table, factored by scale factor (SF). That is, the SF determines the size of the raw data outside the Db (i.e.: SF = 100 means that the sum of all base tables equals 100GB).

To be compliant with TPC-H benchmark, [8] recommends that the Db must be implemented using a commercially available DBMS, with support to queries and refresh functions against all tables on a 7 days of week, 24 hours by day (7x24). The minimum required to run the benchmark holds business data from 10.000 suppliers, with almost ten million rows representing a raw storage capacity of about 1GB (i.e.: SF = 1).

The performance metric reported by TPC-H is called TPC-H Composite Query-per-Hour Performance Metric (QphH), and reflects multiple aspects of the system's capability to process queries. TPC-H benchmark is composed by 22 *ad-hoc* business-oriented queries (16 of which carried from other TPC benchmarks) that include a variety of operators and selectivity

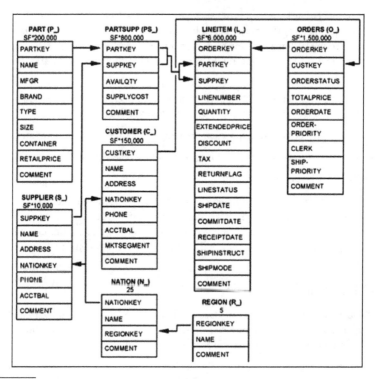

(SOURCE: TPC Benchmark™ H documentation, URL: http://www.tpc.org)

Figure 1. TPC-H database schema.

constraints, whose the objective is to assist the decisions makers in the business analysis (pricing and promotions, supply and demand management, profit and revenue management, customer satisfaction study, market share study and shipping management) [3; 8]. A typical query (Table 1) uses tables "join" and, in most cases, aggregate functions and "group by" clause. The workload of benchmark consists of a data load, the execution of queries in both single and multi-user mode and two refresh functions.

3. Design of experiments overview

In science, the researchers' interest is focused on systems (processes) investigations where, usually, there are several variables for analysis. Often, the investigations are centered on individual changes produced by the variables, as well in their interactions (Figure 2).

Traditionally, in an investigation, experiments are planned to study the effects of a single variable (factor) in a process. However, the combined study of multiple factors represents a

Query	Join	Aggregate functions					Group by	Sub query
		Avg()	Count()	Max()	Min()	Sum()		
Q1		✓	✓			✓	✓	
Q2	✓				✓			✓
Q3	✓					✓	✓	
Q4	✓		✓				✓	✓
Q5	✓					✓	✓	
Q6						✓		
Q7	✓					✓	✓	✓
Q8	✓					✓	✓	✓
Q9	✓					✓	✓	✓
Q10	✓					✓	✓	
Q11	✓					✓	✓	✓
Q12	✓					✓	✓	
Q13	✓		✓				✓	✓
Q14	✓					✓		
Q15	✓			✓			✓	✓
Q16	✓		✓				✓	✓
Q17	✓	✓				✓		✓
Q18	✓					✓	✓	✓
Q19	✓					✓		
Q20	✓							✓
Q21	✓		✓				✓	✓
Q22	✓	✓	✓			✓	✓	✓

Table 1. TPC-H queries' characteristics.

way to determine the main effects, as well as the interaction effects among the factors under-lying the process. The DoE is a framework of statistical techniques, such as the results can produce valid and objective conclusions [9].

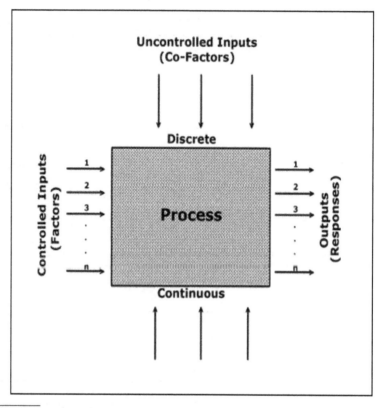

(SOURCE: NIST/Sematech, 2006)

Figure 2. A classical process.

DoE methodology have been used very successfully in the verification, improvement and reducing the processes variability with impacts on costs and development time [9; 10]. A relevant class of DoE techniques is called factorial design, whose goal is to study and analyze the results (effects) produced by multiple variables of a process.

The beginning of a factorial design is a careful selection of a fixed number of levels for each set of factors. The experiments should be performed with all combinations factor/level. For example, if there are l_1 levels to the first variable, l_2 for the second,..., l_k for the k-th factor, the full array of $l_1, l_2,..., l_k$ plays will be classified as factorial design $l_1 \times l_2 \times... \times l_k$.

The default schema for designs with two levels uses the notation "−" (negative) and "+" (positive) to denote the low and high levels of each factor, respectively [9; 10; 11]. For example, a 2x2 factorial design with two factors (X_1 and X_2) and two levels (low and high), requires four experimental plays (Table 2).

Exp.	X_1	X_2	Result
1	–	–	Y_1
2	+	–	Y_2
3	–	+	Y_3
4	+	+	Y_4

Table 2. Experimental matrix.

3.1. Full factorial design

When all combinations of factors are running at the same number of times for each level, the experiment is classified as 2^k *full factorial design*. Thus, the factorial design presented in Table 2 is classified as 2^k ($k = 2$) full factorial design [9; 10]. The most intuitive approach to study such factors would be vary the factors of interest in a full factorial design (trying all possible combinations of settings). For example, a 2^3 full factorial with three factors (X_1, X_2, and X_3) at two levels requires eight experimental plays (Table 3), while to study 5 factors at two levels, the number of runs would be $2^5 = 32$, and $2^6 = 64$, and so on. So, the number of runs required for 2^k full factorial design grows geometrically as k increases, and therefore even the number of factors is small, a full factorial design can become big quickly. In these circumstance, it is recommended [9; 10] to use *fractional factorials designs*.

Exp.	X_1	X_2	X_3	Result
1	–	–	–	Y_1
2	+	–	–	Y_2
3	–	+	–	Y_3
4	+	+	–	Y_4
5	–	–	+	Y_5
6	+	–	+	Y_6
7	–	+	+	Y_7
8	+	+	+	Y_8

Table 3. ($k = 3$) full factorial design.

3.2. Fractional factorial design

Fractional factorial designs represents one way where only a fraction of appropriate combinations required for 2^k full factorial designs is selected for execution. Fractional designs are commonly used when one wants to investigate k factors with smaller number (2^{k-p}) of experiments, where p is the reduction factor [9; 10]. For example, the 2^3 full factorial design (Table 3) can be re-written as a fractional factorial design $2_{III}^{3-1} = \dfrac{2^3}{2} = 4$, where 4 is the number of experimental plays required (Table 4). In the current example, the design is described as 2^{3-1} design of resolution III (three). This means that you study overall $k = 3$ factors, however, $p = 1$ of those factors were generated from the interactions of $2^{[(3-1)-4]}$ full factorial design.

Exp.	X_1	X_2	X_3	Result
1	–	–	+	Y_1
2	+	–	–	Y_2
3	–	+	–	Y_3
4	+	+	+	Y_4

Table 4. ($k = 3, p = 1$) fractional factorial design.

The design does not give full resolution, that is, there are certain interaction effects that are confounded with (identical to) other effects. However, fractional designs requires a smaller number of plays as compared to the full factorial design, but they assumption implicitly that higher-order interactions do not matter. Therefore interactions greater than two-way particularly could escape of detection.

4. Case study

To illustrate the effectiveness of DoE methodology we will apply it in the Db scope, through a case study to know the influence of parameters in the Db performance. This case study will be divided by phases: the first, comprehends a study with a *full factorial design* at two levels, requiring 2^k ($k = 5$), 32 experimental plays. The second phase deals with a *fractional factorial design* 2^{k-p} ($k = 5$ and $p = 1$) resolution V, requiring 16 experimental plays. All proposed experiments will be performed at the same computing environment according to the techniques previously chosen.

Thus, this case study uses the PostgreSQL DBMS (version 8.4.11), through the implementation of a database of 1GB, populated with *dbgen* application (SF = 1) from TPC-H benchmark. Between the 22 queries provided by TPC-H benchmark, we choose to use four queries with a common SQL feature (i.e.: tables "join", aggregate functions and commands to grouping and ordering data) and mostly related to the customer satisfaction study:

- Q10 – identifies customers who might be having problems with the parts that are shipped to them;

- Q13 – seeks relationships between customers and the size of their orders;

- Q18 – ranks customers based on their having placed a large quantity order; and

- Q22 – identifies geographies where there are customers who may be likely to make a purchase.

In this study, the parameters selected (Table 5), intuitively, looks be significant to the Db performance according to the queries characteristics.

	Parameters	Low (–)	High (+)
A	*shared_buffers*	32MB	1024MB
B	*temp_buffers*	8MB	64MB
C	*work_mem*	1MB	1536MB
D	*wall_buffers*	64KB	1MB
E	*effective_cache_size*	128MB	1536MB

Table 5. Design factors.

The PostgreSQL parameters (experiments factors, Table 5) were set according to the suggestions from the PostgreSQL official documentation (http://www.postgres.org), and the values from low level, are standards of installation, while the high level values, were customized according to the computing environment characteristics (a virtual machine implemented over Intel Core™ i5 360 3.20GHz CPU, GNU/Linux i386 openSUSE 11.3, 2GB RAM and hard disc 50 GB). That is, *shared_buffers* (amount of memory reserved for data cache) was set to 50% of total memory; *work_mem* (amount of memory used for sort operations and hash tables) and *effective_cache_size* (amount of memory available for disk cache used by the operating system and Db) were set to 75% of total memory; *temp_buffers* (maximum number of temporary buffers used by each Db session) and *wal_buffers* (useful for systems with high need to write to disk) have 64MB and 1MB, respectively.

4.1. Phase I – 2^k full factorial design

The experiments performed in this phase were structured with five factors at two levels (Table 5), resulting in a 2^5 *full factorial* (32 experimental plays). Each experiment is composed of two replicates and a sample of the experimental matrix, whose the results are the execution time (in seconds) – average time of queries answers 2_{III}^{k-p} – is presented in Table 6. In this table (Table 6), each column contains – (negative) or + (positive) signs to indicate the setting of the respective factor (low or high, respectively). For example, in the first run of the experiment, set all factors A through E to the plus setting, in the second run, set factors A to D to the positive setting, factor D to the negative setting, and so on.

In Table 7 can be found the main effects of factors $\left(\mu = \dfrac{1}{N} \sum_{i=1}^{N} t_i,\ N = 3 \right)$, where E = effect, f = factor [A..E] and Q = query), measured from each query (Q10, Q13, Q18 and Q22). The effects of factors were calculated by the sum of multiplying levels (– and +) by execution time (Y) across all 32 rows. Thus, for query Q10, the effect of factors A and B were estimated as:

Exp.	Factors					Execution time			
	A	B	C	D	E	Q10	Q13	Q18	Q22
1	−	−	−	−	−	0.132	4.122	0.109	0.840
2	−	−	−	−	+	0.126	21.710	0.109	1.067
3	−	−	−	+	−	0.137	4.198	0.113	1.029
4	−	−	−	+	+	0.118	22.052	0.112	1.069
5	−	−	+	−	−	0.185	3.033	0.117	2.553
6	−	−	+	−	+	0.160	3.148	0.119	2.052
7	−	−	+	+	−	0.223	4.097	0.115	1.776
8	−	−	+	+	+	0.169	3.121	0.139	2.857
9	−	+	−	−	−	0.133	4.257	0.116	1.336
10	−	+	−	−	+	0.120	23.552	0.118	1.170
11	−	+	−	+	−	0.132	4.131	0.108	0.830
12	−	+	−	+	+	0.134	27.506	0.113	1.124
13	−	+	+	−	−	0.185	2.992	0.121	2.473
14	−	+	+	−	+	0.163	3.056	0.112	2.223
15	−	+	+	+	−	0.206	3.152	0.127	2.662
16	−	+	+	+	+	0.136	3.373	0.120	1.841
17	+	−	−	−	−	0.336	11.871	0.185	4.062
18	+	−	−	−	+	0.273	14.873	0.147	1.421
19	+	−	−	+	−	0.360	10.997	0.247	4.760
20	+	−	−	+	+	0.265	17.241	0.166	1.451
21	+	−	+	−	−	0.279	7.402	0.151	3.458
22	+	−	+	−	+	0.293	8.605	0.182	5.824
23	+	−	+	+	−	0.305	7.537	0.159	4.149
24	+	−	+	+	+	0.320	8.408	0.154	4.020
25	+	+	−	−	−	0.281	10.905	0.177	5.236
26	+	+	−	−	+	0.287	12.144	0.164	1.380
27	+	+	−	+	−	0.313	11.118	0.195	4.722
28	+	+	−	+	+	0.256	18.553	0.169	1.382
29	+	+	+	−	−	0.272	7.460	0.157	3.746
30	+	+	+	−	+	0.302	7.957	0.154	4.772
31	+	+	+	+	−	0.316	7.488	0.165	3.327
32	+	+	+	+	+	0.307	7.718	0.157	5.582

Table 6. Experimental matrix.

$E_{fi}Q_j$, $i=1..5$; $j=1..4$ Similarly, the same methodology was employed to estimate all others effects of factors.

We use the analysis of variance (ANOVA) to know the influence of factors in the Db performance. According to the ANOVA (Table 8) it appears that the effect of factor A is statistically significant ($p<.05$) for queries Q10, Q18 and Q22, and marginally significant ($p<.01$) for query Q13. It also stands out that the factors C and E are marginally significant for query Q10 and statistically significant for query Q13. However, such factors do not seems to show influence for query Q18. On the other hand, for query Q22 the factor C is statistically significant, while the factor E is marginally significant.

Factors	Q10	Q13	Q18	Q22
A (shared_buffers)	0.144	2.049	0.054	2.024
B (temp_buffers)	0.008	0.184	0.003	0.088
C (work_mem)	0.026	8.168	0.006	1.277
D (wall_buffers)	0.011	0.850	0.007	0.064
E (effective_cache_size)	0.023	6.141	0.008	0.483

Table 7. Effects of factors.

After estimate the effects of each factor and analyze them through the ANOVA, we used a analysis of sensitivity of factors (Table 9), suggested by [5], whose goal is create a rank of the factors. This methodology consists of a *"sorting method, where the effects are normalized with respect to the maximum effect, rounded to the first decimal point, and sorted in descending order"* [5].

For example, the sensitivity effect

$$E_A Q10 = \left| \frac{1}{N}\sum Y_+ - \frac{1}{N}\sum Y_- \right| = \frac{4.764}{16} - \frac{2.460}{16} = 0.298 - 0.154 = 0.144 \right|$$ and

$$E_B Q10 = \left| \frac{3.544}{16} - \frac{3.679}{16} = 0.008 \right|.$$ of factors A and B for query Q10 were estimated as $\left(S_{f_i}Q_j = E_{fi}Q_j / MAX(E_{f_i}Q_j), i=1..5; j=1..4 \right)$ and $S_A Q10 = 0.144/0.144 = 1.0$. All others sensitivity effects were estimated in the same way (Table 8).

Once the sensitivity effect of factors was estimated, they can be rated with respect to its range of influence (Figure 3) based on number of factors studied (i.e.: 5, Table 5). According to this range, each factor has 0.2 units of influence and, therefore such factors with the same normalized effect can be assigned at the same rank. For example, factors with sensitivity effect 0.2 (factor E for query Q10) and 0.3 (factor A for query A13) will be at the same ranking.

The ranking of sensitivity effect of factors is presented in Table 10. These results corroborates with the ANOVA analysis (Table 8) and reveals that, factors statistically significant (i.e.: factor

(a) Q10						(b) Q13				
Factors	SS	df	MS	F	p	Factors	SS	df	MS	F	p
A	0.498	1	0.498	188.789	0.000	A	100.717	1	100.717	4.550	0.036
B	0.002	1	0.002	0.653	0.421	B	0.814	1	0.814	0.037	0.848
C	0.016	1	0.016	6.100	0.015	C	1601.020	1	1601.020	72.331	0.000
D	0.003	1	0.003	1.058	0.306	D	17.343	1	17.343	0.784	0.378
E	0.013	1	0.013	4.821	0.031	E	905.090	1	905.090	40.890	0.000
Error	0.237	90	0.003			Error	1992.105	90	22.135		
Total SS	0.768	95				Total SS	4617.089	95			

(c) Q18						(d) Q22				
Factors	SS	df	MS	F	p	Factors	SS	df	MS	F	p
A	0.069	1	0.069	89.169	0.000	A	98.355	1	98.355	65.706	0.000
B	0.000	1	0.000	0.304	0.583	B	0.188	1	0.188	0.126	0.724
C	0.001	1	0.001	1.192	0.278	C	39.146	1	39.146	26.152	0.000
D	0.001	1	0.001	1.733	0.191	D	0.099	1	0.099	0.066	0.797
E	0.001	1	0.001	1.924	0.169	E	5.592	1	5.592	3.736	0.056
Error	0.070	90	0.001			Error	134.720	90	1.497		
Total SS	0.143	95				Total SS	278.100	95			

Table 8. ANOVA table.

Factors	Q10	Q13	Q18	Q22
A	1.0	0.3	1.0	1.0
B	0.1	0.0	0.1	0.0
C	0.2	1.0	0.1	0.6
D	0.1	0.1	0.1	0.0
E	0.2	0.8	0.1	0.2

Table 9. Sensitivity effect of factors.

| 0.0 |– 0.2 |– 0.4 |– 0.6 |– 0.8 |– 1.0 |

Figure 3. Range of influence of factors.

Factors	Q10	Q13	Q18	Q22
A	1	4	1	1
B	5	5	5	5
C	4	1	5	3
D	5	5	5	5
E	4	2	5	4

Table 10. Sensitivity effect of factors.

A for queries Q10, Q18 and Q22) are the most sensitivity (close to 1.0), while those effects marginally significant (factors C and E for query Q10) have low influence (close to 0.0). Here also these factors do not looks have influence for query Q18.

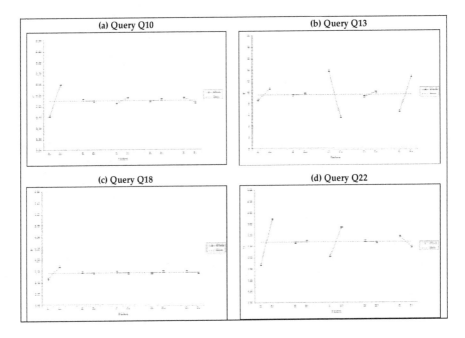

Figure 4. Effect of factors.

Such observations can be highlighted by graphs of the changes of effects versus factors levels (Figure 4). They confirms the hypothesis that the factor A is the most significant for queries Q10, Q18 and Q22. Through a visual inspection, it should be noted that the factors classified with low influence are very close to the average (i.e.: factors B, C, D and E for query Q18). The graphs also confirms that factors C and E are significant for queries Q13 and Q22.

4.2. Phase II – 2^{k-p} fractional factorial design

The second phase of this study case uses a $S_BQ_{10}=0.008/0.144=0.1$ *fractional factorial* (16 experimental plays). Here, it is employed the concept of design resolution, such as the study overall $k = 5$ factors, however, $p = 1$ of those factors were generated from the interactions of a full $2^{[(5-1)-4]}$ factorial design. As result, the design does not give full resolution, that is, there are certain interaction effects that are confounded with other effects (i.e.: factor E, generated as result of one-way interactions between factors A, B, D and D).

As mentioned before, the experiments are composed of two replicates performed at the same computing environment used during the Phase I (Section 4.1). A sample of the experimental matrix with execution time (in seconds) – average time of queries answers – is presented in Table 11.

Exp.	Factors					Execution time			
	A	B	C	D	E	Q10	Q13	Q18	Q22
1	−	−	−	−	+	0.126	21.710	0.109	1.067
2	−	−	−	+	−	0.137	4.198	0.113	1.029
3	−	−	+	−	−	0.185	3.033	0.117	2.553
4	−	−	+	+	+	0.169	3.121	0.139	2.857
5	−	+	−	−	−	0.133	4.257	0.116	1.336
6	−	+	−	+	+	0.134	27.506	0.113	1.124
7	−	+	+	−	+	0.163	3.056	0.112	2.223
8	−	+	+	+	−	0.206	3.152	0.127	2.662
9	+	−	−	−	−	0.336	11.871	0.185	4.062
10	+	−	−	+	+	0.265	17.241	0.166	1.451
11	+	−	+	−	+	0.293	8.605	0.182	5.824
12	+	−	+	+	−	0.305	7.537	0.159	4.149
13	+	+	−	−	+	0.287	12.144	0.164	1.380
14	+	+	−	+	−	0.313	11.118	0.195	4.722
15	+	+	+	−	−	0.272	7.460	0.157	3.746
16	+	+	+	+	+	0.307	7.718	0.157	5.582

Table 11. Experimental matrix.

In Table 12 are presented the effects of factors for each query (Q10, Q13, Q18 and Q22). These effects were calculated with the same methodology used in the Phase I (Section 4.1), but here the sum of multiplying levels (− and +) with execution time (Y) across all 16 rows.

Factors	Q10	Q13	Q18	Q22
A (shared_buffers)	0.140	1.708	0.052	2.008
B (temp_buffers)	0.000	0.113	0.004	0.027
C (work_mem)	0.021	8.295	0.001	1.678
D (wall_buffers)	0.005	1.182	0.003	0.173
E (effective_cache_size)	0.018	6.059	0.003	0.344

Table 12. Effects of factors.

The influence of factors were studied with ANOVA (Table 13). It's noteworthy that factor A is statistically significant for queries Q10, Q18 and Q22, but it does not seems influential for query Q13. By the other hand, the factors C and E are statistically significant for query Q13. We also note that, with the fractional factorial experiments, there are no factors marginally significant for query Q10, so this query, as well as query Q18 have only one factor significant (factor A), while for query Q22 the factors A and C are statistically significant.

(a) Q10						(b) Q13				
Factors	SS	df	MS	F	p	Factors	SS	df	MS	F	p
A	0.237	1	0.237	91.951	0.000	A	34.996	1	34.996	1.283	0.264
B	0.000	1	0.000	0.000	1.000	B	0.154	1	0.154	0.006	0.941
C	0.005	1	0.005	2.016	0.163	C	825.760	1	825.760	30.268	0.000
D	0.000	1	0.000	0.127	0.723	D	16.758	1	16.758	0.614	0.438
E	0.004	1	0.004	1.527	0.223	E	440.609	1	440.609	16.150	0.000
Error	0.108	42	0.003			Error	1145.839	42	27.282		
Total SS	0.354	47				Total SS	2464.116	47			

(c) Q18						(d) Q22				
Factors	SS	df	MS	F	P	Factors	SS	df	MS	F	p
A	0.033	1	0.033	42.577	0.000	A	48.394	1	48.394	31.541	0.000
B	0.000	1	0.000	0.195	0.661	B	0.009	1	0.009	0.006	0.939
C	0.000	1	0.000	0.035	0.853	C	33.790	1	33.790	22.022	0.000
D	0.000	1	0.000	0.170	0.682	D	0.361	1	0.361	0.235	0.630
E	0.000	1	0.000	0.183	0.671	E	1.417	1	1.417	0.924	0.342
Error	0.033	42	0.001			Error	64,.442	42	1.534		
Total SS	0.066	47				Total SS	148.413	47			

Table 13. ANOVA table.

After known the ANOVA results, we also employee the analysis of sensitivity of factors to rank them (Table 14) according to the range of influence (Figure 3). The results reveals that such factors classified as statistically significant by ANOVA (Table 13) (i.e.: factor A for queries Q10, Q18 and Q22) are the most sensitivity. It stands out that A is the only factor that seems influential for queries Q10 and Q18.

Factors	Q10	Q13	Q18	Q22
A	1	4	1	1
B	5	5	5	5
C	5	1	5	2
D	5	5	5	5
E	5	3	5	4

Table 14. Sensitivity effect of factors.

Through the graphs of the changes of effects versus factors levels (Figure 5) it can be noted that the factor A is the most significant for queries Q10, Q18 and Q22. The a visual inspection stands out the factors classified with low influence are very close to the average (i.e.: factors B, C, D and E for queries Q10 and Q18). The graphs also highlights that factors C and E are significant for queries Q13 and Q22.

4.3. Analysis of results

Through the present study were made several experiments using two DoE techniques (2^k full factorial design and 2^{k-p} fractional factorial design) to investigates how different Db parameters can influence in its performance.

The Phase I (Section 4.1) comprehends a study with a 2^k full factorial design ($k = 5$), whose results highlighted the influence of the factors and rated them in concordance with its sensitivity. According to the ANOVA (Table 8) there are factors statistically significant for one query, but marginally significant for others. So, we employed the analysis of sensitivity (Table 10), that corroborated with the ANOVA results. Thus, according to the queries characteristics used in this case study, the results suggests the factor A as the most significant, followed by factors C and E rated as very significant and significant, respectively, while the others (factors B and D) looks have low influence in the Db performance.

In the Phase II (Section 4.2) a same research was conducted, but using a 2^{k-p} fractional factorial design ($k = 5$, $p = 1$). With the fractional design we come to the results with half of the work required by full design and, through them, we also know the influence of each factor (Table 14). The results were similar to those succeeded before (Section 4.1) and rated the factor A as the most significant, followed by factors C and E as very significant and significant, respectively, while the others (factors B and D) with low influence in the Db performance.

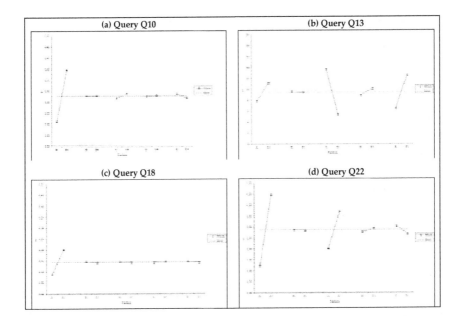

Figure 5. Effect of factors.

To perform comparison of results between both techniques, it's noteworthy that even with similar results, full design is more accurate. For example, it stands out such factors classified as marginally significant by ANOVA (Table 8) from full design (i.e.: C and E for query Q10), do not appear in fractional design. However, despite the accuracy, the full design is more laborious and, therefore should require more resources (depending on the number of factors). Anyway, we could state that, in this study, both techniques proved to be effective for identification and classification of influential factors (parameters) to the Db performance.

In this study, we assume that all queries have same importance. So intuitively it seems that factor A (*shared_buffers*, amount of memory reserved for data cache) is one of the most significant to improve the Db performance, as it appears as rated first for three queries out of five with both techniques chosen. The factor C (*work_mem*, amount of memory used for sort operations and hash tables) also seems very significant, as it was rated as first and second for two queries (Q13 and Q22, respectively), while the factor E (*effective_cache_size*, amount of memory available for disk cache used by the operating system and Db) seems marginally significant. Since their rate can vary according to the query. Another interesting observation is that factors B and D (*temp_buffers*, related to the maximum number of temporary buffers used by each Db session, and *wall_buffers*, useful for systems with high need to write to disk, respectively) never seems important for the individual queries. Therefore, the results suggests that these two parameters should have low impact to improve the Db performance. All results are summarized in Table 15.

Influence	Parameter
High	shared_buffers
	work_mem
Medium	*effective_cache_size*
Low	temp_buffers
	wall_buffers

Table 15. Parameters influence.

5. Related work

A quick search on the contemporary literature reveals some works addressing to the use of DoE in the several scopes. At the computer science area, the use of DoE is explored by [12] through a comprehensive study about techniques for software engineering experimentation. Other works [13; 14] are devoted to the algorithmic optimizations by means of DoE.

There are also a lot of works approaching the Db performance subject. For example, [15] approaches the optimizations of Db systems through a statement of a new problem, that is the Web-based interactive applications. [16] report a performance study with different Db architectures and provide useful information for the enterprise architects and database administrator in determining the appropriate Db architecture. Techniques to automate the setting of tuning parameters in specifics software applications could be found in [17], as well as in [18]. The importance of best practices and the database administrator knowledge for autonomic Db tuning is pointed by [19]. In [20] is introduced a algorithm to select a small subset of Db statistics, such that it can improve the benefits over maintaining base-table statistics. To [21] the challenge in making Db systems truly self-tuning is a tall task, due the different abstractions for workloads and different constraints on the desired solution (i.e.: the complexity of internal components of the Db architecture). In [22] is discussed a way to avoid the trial and error process of SQL tuning, through by choosing a small set of experiments in order to reach a satisfactory plan to execute queries.

The o use of DoE techniques is formally explored in the Db scope. In the [5; 6], the database performance evaluation was studied by a statistical approached. The authors define a statistical methodology, which may be useful to minimize the effort related with database tuning activities. Following in this line, the study presented by [7] describes a software application, whose purpose is to automate the database parameters configuration by means of DoE.

In summary, in the Db performance area there are much of the effort to take the tests and results comparison, however only a little portion of the studies uses the DoE methodology to planning and analysis of the experiments.

6. Final considerations

This chapter presented a study to investigate the influence of Db parameters in its performance. Two DoE techniques (full factorial design and fractional factorial design) were applied in a case study with PostgreSQL DBMS and TPC-H benchmark, to assist in the investigations of how each parameter may influence in the Db performance.

To the case study, were selected five parameters mostly related with the Db tuning and conducted different experiments with the DoE techniques chosen. At the Phase I, we studied the influence of parameters through a 2^k full factorial design, whose the analysis suggests, with high degree of confidence, the parameter shared_buffer as the most significant, while work_mem and effective_cache_size can be classified as very significant and significant, respectively. According to the analysis, the others parameters (temp_buffers and wall_buffers) looks have low influence in the Db performance. The Phase II comprehended a similar study, but using 2^{k-p} fraction factorial design. The results were very similar with those suggested in Phase I, that is the shared_buffer, work_mem and effective_cache_size looks have influence in the Db performance, while the others not. It stands out in Phase II, that those parameters marginally significant with full design, do not appear in fractional design. This characteristics leading us to conclude that, although being the most laborious, full design is more accurate. But, by the other hand, according to the analysis of case study results, it is also feasible to reach the same conclusions with fractional design using half of the work required by the full design.

It should be noted that Db technology was used in this study as a vehicle to demonstrate how the DoE methodology can help in the design of experiments and its analysis and used as tool in several scopes, like in the computing science field. We also emphasizes that this study did not aim to close the subject about the use of DoE in the computing scope, instead it we sought disclose the effectiveness of this methodology applied in this context. Thus, we can conclude that DoE methodology is a promising to assist in quantitative analysis, for example in the investigation of influential parameters in the Db performance.

Author details

Eduardo Batista de Moraes Barbosa[1*] and Messias Borges Silva[1,2]

*Address all correspondence to: ebmb@yahoo.com

1 São Paulo State University "Júlio de Mesquita Filho" – UNESP, School of Engineering at Guaratinguetá – FEG, Guaratinguetá, SP, Brazil

2 University of São Paulo – USP, School of Engineering at Lorena – EEL, Estrada Municipal do Campinho, s/n – Lorena, SP, Brazil

References

[1] Gray, J. (Ed.) The Benchmark Handbook for Database and Transaction Systems. 2nd. Edition. Morgan Kaufmann, 1993.

[2] Leutenegger, S. T.; Dias, D. A Modeling Study of the TPC-C Benchmark. ACM SIG-MOD Record, Vol. 22, Issue 2, p. 22-31, 1993.

[3] Poess, M., Floyd, C. New TPC Benchmarks for Decision Support and Web Commerce. ACM SIGMOD Vol. 29 Issue 4, p. 64-71, 2000.

[4] Hsu, W. W.; Smith, A. J.; Young. H. C. I/O Reference Behavior of Production Database Workloads and the TPC Benchmarks – An Analysis at the Logical Level. ACM Transactions on Database Systems (TODS), Vol. 26, Issue 1, p. 96-143, 2001

[5] Debnath, B. K.; Lilja, D. J.; Mokbel, M. F. Sard: A Statistical Approach for Ranking Database Tuning Parameters. In: IEEE 24th International Conference on Data Engineering Workshop (ICDEW), pp. 11-18, 2008.

[6] Debnath, B. K.; Mokbel, M. F.; Lilja, D. J. Exploiting the Impact of Database System Configuration Parameters: A Design of Experiments Approach. In: IEEE Data Eng. Bull., pp. 3-10, 2008.

[7] Duan, S.; Thummala, V.; Babu, S. Tuning Database Configuration Parameters with iTuned. In: Proceedings of Very Large Database Endowment (PVLDB), pp. 1246-125, 2009.

[8] TPC Benchmark™ H (Decision Support) Standard Specification Revision 2.8.0. URL: http:// www.tpc.org (Accessed: Dec., 2012).

[9] NIST/SEMATECH. e-Handbook of Statistical Methods. 2006. URL: http:// www.itl.nist.gov/ div898/handbook. (Accessed: Dec., 2012)

[10] Montgomery, D. C. Design and Analysis of Experiments (5th Edition). John Wiley & Sons Inc., 2001.

[11] Box, G. E. P.; Hunter, W. G.; Hunter, J. S. Statistics for Experimenters (1 Ed.). John Wiley & Sons, 1978.

[12] Juristo, N.; Moreno, A. M. Basics of Software Engineering Experimentation. Springer, 2001.

[13] Adeson-Diaz, B.; Laguna, M. Fine-Tuning of Algorithms Using Fractional Experimental Designs and Local Search. Operations Research, Vol. 54, No. 1, p. 99-114, 2006.

[14] Arin, A.; Rabadi, G.; UNAL, R. Comparative studies on design of experiments for tuning parameters in a genetic algorithm for a scheduling problem. Int. Journal of Experimental Design and Process Optimisation, Vol. 2, N. 2, p. 103-124, 2011.

[15] Florescu, D., Kossmann, D. Rethinking Cost and Performance of Database Systems. ACM SIGMOD Record, Vol. 38, No. 1, p. 43–48, 2009.

[16] Chen, S., Ng, A., Greenfield. P. Concurrency and Computation: Practice and Experience. Concurrency Computat.: Pract. Exper. doi: 10.1002/cpe.2891. 2012.

[17] Stillger, M.; Lohman, G.; Markl, V.; Kandil, M. Leo – DB2's Learning Optimizer. In: Proceedings of Very Large Database Endowment (PVLDB), p. 19-28, 2001.

[18] Badu, S. Towards Automatic Optimization of MapReduce Programs. In: Proceedings of the 1st ACM symposium on Cloud computing (SoCC '10), p. 137-142, 2010.

[19] Wiese, D.; Rabinovitch, G; Reichert, M; Arenswald, S. Autonomic Tuning Expert – A framework for best-practice oriented autonomic database tuning. In: Proceedings of the 2008 conference of the center for advanced studies on collaborative research: meeting of minds (CASCON '08), p. 3:27-3:41, 2008.

[20] Bruno, N., Chaudhuri, S. Exploiting Statistics on Query Expressions for Optimization. In: Proceedings of the 2002 ACM SIGMOD international conference on Management of data, p. 263-274, 2002.

[21] Chaudhuri, S., Narasaaya, V. Self-Tuning Database Systems: A Decade of Progress. In: Proceedings of the 33rd international conference on Very large data bases, p. 3-14, 2007.

[22] Herodotou, H.; Badu, S. Automated SQL Tuning through Trial and (sometimes) Error. In: Proceedings of the Second International Workshop on Testing Database Systems (DBTest '09), article No. 12, 2009.

Chapter 5

Taguchi Method Applied to Environmental Engineering

Ana Paula Barbosa Rodrigues de Freitas, Leandro Valim de Freitas,
Carla Cristina Almeida Loures, Marco Aurélio Reis dos Santos,
Geisylene Diniz Ricardo, Fernando Augusto Silva Marins,
Hilton Túlio Lima dos Santos, Gisella Lamas Samanamud,
Mateus Souza Amaral and Messias Borges Silva

Additional information is available at the end of the chapter

1. Introduction

Over the last decades, environmental concerns have become more critical and frequent. This is, mainly, due to population growth and the increase of industrial activities in which anthropogenic actions have reached catastrophic proportions resulting in changes of soil, air, and water quality [1].

Environmental pollution by industrial effluent is being characterized as one of the major causes of the aggravation of this problem. Residues, in general, produce diversified compounds, containing, frequently, pollutants that are toxic and resistant to conventional treatments such as coagulation/flocculation or biodegradation [2], and they are eventually discharged, in most of the cases, in an inadequate way causing severe damages. Regarding the environmental problem, researchers were driven to study the feasibility of new techniques and methodologies, as well as, the emission and pollutant discharge control. In order to apply the pollution control and to attend environmental legislation, patterns and quality indicators were established. In terms of water quality: oxygen concentration, phenols, Hg, pH, temperature, among other requirements [3].

Companies search for new environmental alternatives to treat generated residuals. The environmental reality is demanding for further actions to mitigate industrial impacts on water. Therefore, water treatment has become a mandatory investment to industries, institutions, and others with the aim to attending environmental laws, as well as ISO 14000 series.

In this context, textile sector can be referred because of its great industrial area that generates a high volume of effluent deeply colored and containing high concentration of organic compounds, which if not treated, may cause serious damage to environmental contamination [4].

Hazardous waste treatment and the presence of organic pollutants in water have increased the use of alternatives to environmental matrixes such as the use of Advanced Oxidation Processes (AOPs) to residual water treatment [5].

This work features the application of Design of Experiments; Taguchi L_9 Orthogonal Array; in the effluent treatment of polyester resin that is originated from textile industries and through the application of Advanced Oxidation Processes (Heterogeneous Photocatalysis - UV/TiO_2) in the study of chemical oxygen demand.

2. Advanced oxidation processes

Advanced Oxidation Processes (AOP's) and electrochemical methodologies are developed to treat the contaminants of drinking water and industrial effluents. The oxidation processes are based on reactive species generation that degrades a great variety of organic pollutants, in a quick and non-selective way. Reactive species are unstable and must be generated continuously "in situ", through chemical or photochemical reactions [6].

AOPs are defined as processes with great capacity of producing hydroxyl radicals ($•OH$), that are reactive species. The high standard potential of radical's reduction is demonstrated by Equation 1. This radical is capable of oxidizing a great variety of organic compounds to CO_2, H_2O and inorganic ions originated from heteroatoms [7-8].

$$•OH + e^- + H+ \rightarrow H_2O \qquad\qquad E_o = 2,730 \ V \qquad\qquad (1)$$

Its destructive process is one of its great advantages. Contaminants are chemically destroyed instead of undergoing a phase change that happens, for instance, in physical-chemical processes of adsorption, filtration, precipitation, coagulation, flocculation, sedimentation, flotation, membrane use, organic and inorganic adsorption, centrifugation, reverse osmosis, extraction, distillation and evaporation [9]. The final disposition of solid phases continues being a problem without any solution; therefore, a passive agent [10]. This reagent is very few selective, electrophilic character, easy to produce and detains kinetic reaction control [11].

The hydroxyl radicals can be obtained from strong oxidants, as H_2O_2 and O_3 combined or not with UV radiation, with salts of Iron II or III, combined or not, with radiation, photocatalysis with TiO_2 or water photolosys with UV radiation [12].

The organic matter (OM) present in the system is attacked by hydroxyl radical at the moment that it is generated, and as a consequence of this process, the effluent is degraded to other intermediate products described in Equation 2[13].

$$\bullet OH + OM \rightarrow \text{Intermediates} \tag{2}$$

The several AOPs are split into two groups: Homogenous Processes and Heterogeneous Processes. The former occur in one single phase and use ozone, H_2O_2 or Fenton reagent (mixture of H_2O_2 with salts of Fe^{2+}) as hydroxyl radical generators. The latter uses semiconductors as catalysts (titanium dioxide, zinc oxide, etc.)[14].

3. Heterogeneous photocatalysis

Practical studies using TiO_2 have been developed; however, the reaction mechanism is not totally understandable, yet. Nevertheless, most of researchers agree on some mechanisms steps such as: the excitation of semiconductors species and the formation of h+BV and e-BC, the recombination process among them, O_2, H_2O adsorption and organic species on the semiconductor surface, "trapping" where chemical species donate or accept a pair of electrons e-/h+ preventing the recombination. It is believed that O2 is the main responsible specie to give continuity to the reactions started during the photo-oxidation process, reacting as a formed organic radical and promoting a complete mineralization [15-16]. Figure 1 shows the excitation scheme of the semiconductors.

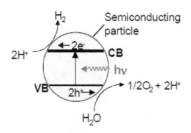

Photocatalytic H_2 Production

Figure 1. Schematic illustration of electricity generation and hydrogen production by solar energy conversion using semiconducting materials. CB: Conduction band, VB: Valence band Source: [17].

Direct oxidation process occurs when a photogerated gap in the valence band of semiconductor reacts directly with the organic compound (Equation 3) [18].

$$R_1(\text{ads}) + \text{h+BV} \rightarrow R_1(\text{ads}) \tag{3}$$

Indirect oxidation process occurs when a photogerated gap in the valence band of semiconductor reacts with H_2O molecule adsorbed on the semiconductor surface producing hydroxyl radical that will oxide the organic material (Equation 4) [18-19].

$$\bullet OH + R_1 \rightarrow R_2 \tag{4}$$

Photocatalytic process has been efficiently used to degrade innumerable recalcitrant substances prior to the biological treatment.

4. Benefits of advanced oxidation processes

Advanced Oxidation Processes offer several advantages when compared to conventional oxidation processes [20-21].

- Assimilate a large variety of organic compounds;
- Complete mineralization of pollutants;
- Destroy resistant refractory compounds to other treatments, as for example, biological treatment;
- May be used in other processes as a pre or post-treatment;
- Used in effluents with high toxicity that can entail a certain difficulty in the biological process treatment;
- Enable *in situ* treatment;
- Do not create reaction by-products;
- Improve the organoleptic properties of the treated water;
- Contain oxidizing power with elevated kinetic reaction.

5. Chemical Oxygen Demand (COD)

Chemical Oxygen Demand (COD) measures the amount of oxygen consumed through the organic material in water and, also represents an essential parameter in the characterization study of sanitary wastewater and industrial effluents. COD is crucial when used along with BOD to analyze and evaluate wastewater biodegradability [22].

By determining COD, the oxidation-reduction reaction is performed in a closed system using potassium dichromate due to its high oxidative capacity and to its application in a large variety of samples and operational feasibility [23].

Sample results of COD using potassium dichromate as an oxidation agent are superior to BOD, because the high oxidative power of the potassium dichromate is greater if compared to the action of micro-organisms, except in rarely cases, as aromatic hydrocarbonetes and pyridine. BOD measures only the biodegradable fraction. The more this value approximates to COD more easily biodegradable the effluent is [22, 24-25].

6. Design of Experiments (DOE)

Design of Experiments has been widely used to optimize processes parameters and to improve the quality of products with the application of engineering concepts and statistics [26].

Design of Experiments is defined as a set of applied statistical planning techniques, conducting, analyzing and interpreting controlled tests with the aim to find and define factors that may influence values of a parameter or of a group of parameters [27].

DOE considers interaction among variables and may be used to optimize operational parameters in multivariable systems [28].

According to [29], design of experiments was studied as a relevant mathematical tool in the area of Advanced Oxidation Processes. Taguchi's Orthogonal Array L_9 was used in this work for the degradation of organic material of the polyester resin effluent and the percentage reduction of the total organic carbon obtained in the treatment was 39.489%. This removal of organic load corresponds to an average ratio of TOC removal. This condition is inclusive of the weight ratio of hydrogen peroxide at 183g, pH = 3, TiO_2 = 0,250 g/l and lamp intensity = 21 W.

Design of experiments was used by [30] in the degradation of organic material of the polyester resin effluent by advanced oxidation processes. Taguchi's Orthogonal Array L_{16} was used to select statistically the most significant factors in the process; being optimized, lately. It was concluded that more influent variables permitted a reduction of 34% COD of the polyester resin effluent.

7. Taguchi method

According to [31], Taguchi's method is a powerful mathematical tool capable to find significant parameters of an ideal process through multiple qualitative aspects.

The application of Taguchi's method [32] consists of:

• Selecting the variable response to be optimized;

• Identifying factors (entry variables) and choosing the levels;

• Selecting the appropriate orthogonal array according to literature [33];

• Performing random experiments to avoid systematic errors;

• Analyzing results by using signal-to-noise ratio (S/N) and analysis of variance (ANOVA);

• Finding the best parameter settings.

There are independent variables or entries in the signal-to-noise ratio that compromises the performance of a process. For this reason, two categories are defined: controllable and non-controllable factors [34].

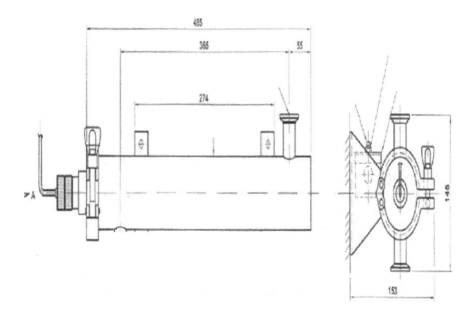

Figure 2. Tubular reactor used for photochemical treatment Source: [28]

Taguchi's method uses orthogonal arrays to study diverse factors with a reduced number of experiments [35]. Besides that, the method can offer other advantages as: process variability reduction, conformity of the expected result and, consequently, operational cost reduction [36].

The analysis of variance (ANOVA) is applied to Taguchi's statistical method to evaluate the significance of parameters used in the process [37].

8. Materials

The polyester resin effluent was conditioned in a chamber at 4 ºC. The oxidation reaction of the effluent was performed in a tubular reactor of Germetec brand, Model GPJ-463/1, with nominal volume of approximately 1L, receiving radiation from a low-pressure mercury lamp type GPH-463T5L emitting UV radiation of 254 mm intensity of 15 W and 21 W, protected by a quartz pipe according to Figure 2.

The design of experiments followed these steps:

- 1L of effluent for 2L of distilled water was firstly added,
- Then, it was added TiO_2,
- The system for the effluent recirculation was turned on,

- H_2O_2 was added, and

- Simultaneously, the UV lamp was turned on.

9. Results and discussion

The design of experiments of polyester resin effluent was performed in a Taguchi's Orthogonal Array L_9. pH, titanium dioxide (TiO_2), ultraviolet lamp and hydrogen peroxide with concentration of 30% w/w were used in this process as controlled variables. Table 1 shows the variables and levels used in the degradation process.

Table 2 shows Taguchi's Orthogonal Array L_9, where experimental procedures were performed at random and, after each experimental procedure, chemical oxygen demand analysis were performed on each experimental condition.

Controlled Variables (Factors)	Level 1	Level 2	Level 3
A- Ph	3.0	5.0	7.0
B- TiO_2 [g/L]	0.083	0.167	0.25
C- H_2O_2 [g]	120.0	151.0	182.0
D- UV [W]	Without	15	21

Table 1. Controlled Variables and Levels

Experiment	pH Factor A	TiO_2 Factor B	H_2O_2 Factor C	UV Factor D
1	1	1	1	1
2	1	2	2	2
3	1	3	3	3
4	2	1	2	3
5	2	2	3	1
6	2	3	1	2
7	3	1	3	2
8	3	2	1	3
9	3	3	2	1

Table 2. Taguchi's Orthogonal Array L_9, with 4 factors and 3 levels each

The COD of the effluent sample *in natura* was initially calculated with a mean value of 49990mg/L and, lately submitted to a pre-treatment. For each experiment, the COD of each sample *in natura* was calculated to an equal period of 60-minute-reaction. Then, the percentage of reduction of COD was calculated for each experiment and the results are shown in Table 3 and 4.

Experiment	pH Factor A	TiO$_2$ Factor B	H$_2$O$_2$ Factor C	UV Factor D	Replica 1 reduction of chemical oxygen demand (%)
1	1	1	1	1	73.44
2	1	2	2	2	78.39
3	1	3	3	3	72.77
4	2	1	2	3	82.35
5	2	2	3	1	69.81
6	2	3	1	2	80.70
7	3	1	3	2	71.13
8	3	2	1	3	78.71
9	3	3	2	1	71.13

Table 3. Result of replica 1 - percentage reduction obtained by experiments for an initial COD amount of 49990 mg/l.

Experiment	pH Factor A	TiO$_2$ Factor B	H$_2$O$_2$ Factor C	UV Factor D	Replica 2 reduction of chemical oxygen demand (%)
1	1	1	1	1	73.11
2	1	2	2	2	76.08
3	1	3	3	3	70.14
4	2	1	2	3	81.69
5	2	2	3	1	65.85
6	2	3	1	2	83.34
7	3	1	3	2	69.15
8	3	2	1	3	81.69
9	3	3	2	1	73.77

Table 4. Results of replica 2 – percentage reduction obtained by experiments, for an initial COD amount of 49990 mg/l.

For experiments performed in the first replica, it is noticeable that oxidative processes reduced COD until 82.345% of the initial amount, being experiment 4 the one with best experimental condition for the degradation experiment, consisting of pH of 5, titanium dioxide of 0.083 g/L, hydrogen peroxide of 151 g and ultraviolet lamp intensity of 21 watts.

The design experiments of the second replica featured that advanced oxidation processes reduced COD until 83.34% of the initial amount, being experiment 6 the one with best experimental variables conditions at pH of 5, titanium dioxide of 0.25g/L, titanium peroxide of 120 g and ultraviolet lamp intensity of 15 Watts.

Factors	Sum of Squares	DF	Mean Sum	Fisher Test	P-Value
Intercept	101741.1	1	101741.1	33172.66	0.000000
pH	40.2	2	20.1	6.56	0.017495
TiO_2	0.2	2	0.1	0.03	0.975085
H_2O_2	264.7	2	132.3	43.15	0.000024
UV	149.6	2	74.8	24.39	0.000232
Error	27.6	9	3.1		

Table 5. Analysis of Variance for polyester resin effluent degradation

Table 5 shows ANOVA factor involved in the polyester resin effluent treatment with the Heterogeneous Photocatalytic Process. The analysis of variance with 95% trust, critical F equal to 4.26 and p-value lower than 5% demonstrated that hydrogen peroxide (F= 6.56 and P-value= 1,7%), temperature (F= 43,15 and P-value= 0,0024%) and lamp intensity(F= 24,39 and P-value= 0,0232%), were significant in the COD removal process.

Taguchi's L_9 statistical design of experiment (Figure 3) showed more significant parameters for the organic material degradation of the effluent, corresponding to pH=5 adjusted to medium level, TiO_2 adjusted to any level, H_2O_2 concentration = 120g and the ultraviolet lamp intensity adjusted to maximum level of 21 W.

According to [38], the influence of peroxide and temperature is related to the efficiency ratio in the use of this compound and its accelerated decomposition in the reactional medium.

Figure 4 shows the most significant factors to a percentage reduction of Chemical Oxygen Demand. The graph of surface response shows an increase in the degradation of polyester resin compounds by the increase in the UV lamp intensity for lower hydrogen peroxide ratio. The highest percentage reduction is of 83%.

10. Conclusions

Advanced Oxidation Process (Heterogeneous Photocatalysis) for the Taguchi design of this work was evaluated, in which values are found to be significant to the chemical oxygen

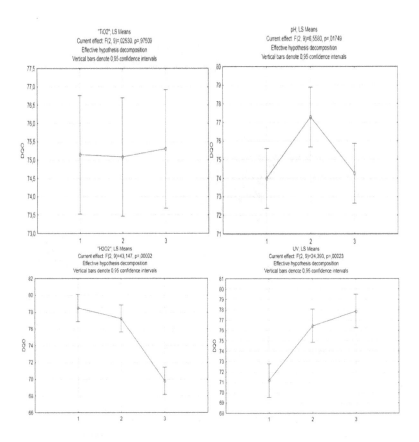

Figure 3. Graphic of Average Impact Factors

demand removal. This design of experiment verified that the highest COD reduction is related to the increase in peroxide hydrogen concentration of 120g, pH=5 and use of UV lamp, since the mechanism of photocatalysis requires energy to the degradation of organic matter of the effluent. Taguchi L_9 Orthogonal Array found was 83%, which demonstrates efficiency on the use of design of experiments and alternative methodologies to the degradation of organic load of polyester resin effluents.

Acknowledgements

The authors would like to thank Fundação para o Desenvolvimento Científico e Tecnológico (FDCT) for translation support.

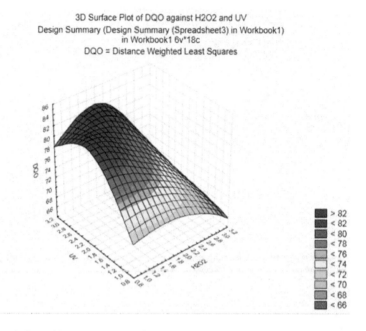

Figure 4. Graphic of influential factors on COD removal

Author details

Ana Paula Barbosa Rodrigues de Freitas[2,3], Leandro Valim de Freitas[1,2],
Carla Cristina Almeida Loures[2,3], Marco Aurélio Reis dos Santos[2], Geisylene Diniz Ricardo[2],
Fernando Augusto Silva Marins[2], Hilton Túlio Lima dos Santos[3], Gisella Lamas Samanamud[4],
Mateus Souza Amaral[3] and Messias Borges Silva[2,3]

1 Petróleo Brasileiro SA (PETROBRAS), Brasil

2 São Paulo State University (UNESP), Brasil

3 University of São Paulo (USP), Brasil

4 University of Texas at San Antonio (UTSA), Brasil

References

[1] Mulbry, W.; Kondrad, S.; Pizarro, C.; Kebede-Westhead, E. Treatment of dairy manure effluent using freshwater algae: Algal productivity and recovery of manure nu-

trients using pilot-scale algal turf scrubbers. Bioresource Technology, v. 99, p. 8137 – 8142, 2008.

[2] Al-Momani, F.; Touraud, E.; Degorce-Dumas, J.R.; ROUSSY, J.; THOMAS, O. Biodegradability enhancement of texile dyes and textile wastewater by VUV photolysis. Journal of Photochemistry and Photobiology, A: Chemistry, v. 153, p. 191– 197, 2007.

[3] Braga, B.; Hespanhol, I.; Lotufo, C.; Conejo, J. G.; Mierzwa, J. C.; De Barros, M. T. L.; Spencer, M.; Porto, M.; Nucci, N.; Juliano, N.; Eiger, S. Introdução à Engenharia Ambiental. 2ª ed. São Paulo: Pearson Prentice Hall, 2005.

[4] Pelegrini, R., Peralta-Zamora, P., Andrade, A.R., Reyes, J., Duran, N., Appl. Catal. B: Envorn.22 (1999) 83.

[5] Segura,Y.; Molina, R.; Martínez, F.; Melero, J. A. Integrated heterogeneous sono–photo Fenton processes for the degradation of phenolic aqueous solutions. Ultrasonics Sonochemistry, v. 16, p. 417–424, 2009.

[6] Oliver, J. H.; Hynook, K.; Pen-Chi, C. Decolorization of wastewater, Crit. Rev.Environmental Science Technology, v. 30, n.4, p. 499–505, 2000.

[7] Poulopoulos, S.G.; Arvanitakis, F.; Philippopoulos, C.J. Photochemical treatment of phenol aqueous solutions using ultraviolet radiation and hydrogen peroxide. Journal of Hazardous Materials, v. 129, p. 64–68, 2006.

[8] Kusic, H.; Koprivanac, N.; Srsan, L. Azo dye degradation using Fenton type processes assisted by UV irradiation: A kinetic study. Journal of Photochemistry and Photobiology A: Chemistry, v. 181 p. 195–202, 2007.

[9] Teixeira, C.P.A.B.; Jardim, W.F. Processos Oxidativos Avançados – Conceitos Teóricos. Campinas: Instituto de Química (IQ) e Laboratório de Química Ambiental (LQA). Universidade de Campinas. (Caderno Temático, v. 3, 2004).

[10] Kunz, A., Peralta-Zamora, P. de Moraes, S. G., Duran, N., Novas Tendências no Tratamento de Efluentes Têxteis, Química Nova. v. 25, n. 1, p. 78-82, 2002. Disponível em: http://www.scielo.br/pdf/qn/v25n1/10428.pdf Acesso em: 04/jan/2013.

[11] Oppenlander, T. Photochemical Purification of Water and Air. Weinheim: Wiley-Vch Verlag, Germany, 2003.

[12] Ferreira, I. V. L.; Daniel, L. A. Fotocatálise heterogênea com TiO2 aplicada no tratamento de esgoto sanitário secundário. Engenharia Sanitária Ambiental, v. 9, n. 4, p. 335-342, 2004.

[13] Khataee, A. R.; Vatanpour, V.; Ghadim, A. R. A. Decolorization of C.I. Acid Blue 9 solution by UV/Nano-TiO2, Fenton, Fenton-like, electro-Fenton and electrocoagulation processes: A comparative study. Journal of Hazardous Materials, v. 161, p. 1225– 1233, 2009.

[14] Tobaldi, D. M.; Tucci, A.; Camera-Roda, G.; Baldi d, G.; Esposito, L. Photocatalytic activity for exposed building materials. Journal of the European Ceramic Society, v. 28, p. 2645–2652, 2008.

[15] Choi, W.; Hoffmann, M. R. Novel Photocatalytic Mechanisms of CHCl3, CHBr3, and CCl3CO2- Degradation and the Fate of Photogenerated Trihalomethyl Radicals on TiO2. Environmental Science Technology, v. 31, p. 89 – 95, 1997.

[16] Chen, J.; Ollis, D. F.; Rulkens, W. H.; Bruning, H. Photocatalyzed oxidation of alcohols and organochlorides in the presence of native TiO2 and metalized TiO2 suspensions. Part (II): Photocatalytic mechanisms. Water Research, v. 33, n. 3, p. 669 – 676, 1999.

[17] López, C. M.; Choi, K.-S. "Enhancement of electrochemical and photoelectrochemica properties of Fibrous Zn and ZnO electrodes" Chem. Commun. 2005, 3328-3330.

[18] Hoffmann, M.; Martin, S.T.; Choi, W; Bahnemann. Chem. Rev., v. 95, p. 69, 1995.

[19] Linsebigler, A. L.; Lu, G.; Yates Jr, J. T. Chem. Rev., v. 95, p. 735, 1995.

[20] Gabardo Filho,H. Estudo e projeto de reatores fotoquímicos para o tratamento de efluentes químicos. 2005. Dissertação (Mestrado em Engenharia Química) –Faculdade de Engenharia Química, Universidade Estadual de Campinas, Campinas.

[21] Domènech, X.; Jardim, W. F.; Litter, M. I. Procesos avanzados de oxidación para La eliminación de contaminantes. In: Eliminacion de Contaminantes. La Plata: Rede CYTED, 2001. Cap. 1.

[22] HU, Z.; GRASSO, D. Chemical Oxygen Demand. In: Water Analysis. 2nd ed. New York: Elsevier Academic Press, 2005. p. 325-330.

[23] Peixoto, A. L. C.; Brito, R. A.; Salazar, R. F. S.; Guimarães, O. L. C.; Izário Filho, H. J. Predição da demanda química de oxigênio no chorume maduro contend reagents de fenton, por meio de modelo matemático emprírico gerado com planejamento factorial complete. Química Nova, v. 31, n.7, p. 1641-1647, 2008.

[24] Piveli, R. P.; Morita, D.M. Caracterização de águas Residuárias / Sólidos. São Paulo: Escola Politécnica/USP, 1998, 11p. (apostila).

[25] Aquino, S. F.; Silva, S. Q.; Chernicharo, C. A. L. Considerações práticas sobre o teste de demanda química de oxigênio (DQO) aplicado a análise de efluentes anaeróbicos (nota técnica). Engenharia Sanitária e Ambiental, v. 11, p. 295-304, 2006.

[26] Wang, T. Y., Huang, C.Y. Improving forecasting performance by employing the Taguchi method. European Journal of Operational Research, p. 1052-1065, 2007.

[27] Bruns, R. E.; Neto, B. B.; Scarminio, I. S. Como Fazer Experimentos. 4. ed. Porto Alegre: Editora Artmed, 2010. 401p.

[28] Ay, F.; Catalkaya, E.C.; Kargi, F. A statistical experiment design approach for advanced oxidation of Direct Red azo-dye by photo-Fenton treatment. Journal of Hazardous Materials, v. 162, p. 230-236, 2009.

[29] Freitas, A. P. B. R. de.; Freitas, L. V. de.; Samanamud, G. L.; Marins, F. A. S.; Loures, C. C. A.; Salman, F.; Santos, H. T. L. dos.; Silva, M. B. Book: Multivariate Analysis in Advanced Oxidation Process. Ed:Intech, p. 73-90, 2012.

[30] Carneiro, L. M. Utilização de Processos Oxidativos Avançados na degradação de efluentes provenientes das indústrias de tintas. 2007, 118 f. Dissertação (Mestrado em Engenharia química) – Escola de Engenharia de Lorena, Universidade de São Paulo, Lorena 2007.

[31] Chiang, Y. M.; Hsieh, H.H. The use of the Taguchi method with grey relational analysis to optimize the thin-film sputtering process with multiple quality characteristic in color filter manufacturing. Computers & Industrial Engineering, v. 56, p. 648-661, 2009.

[32] Barrado, E.; Vega, M.; Grande, P.; Del Valle, J.L. Optimization of a purification method for metal-containing wastewater by use of a Taguchi experimental design. Water Research, v. 30, p. 2309–2314, 1996.

[33] Taguchi, G.; Konish, S. Taguchi Methods: Orthogonal Arrays and Linear Graphs. American Supplier Institute, 1987.

[34] Yang, H.J.; Hwang, P.J.; Lee, S.H. A study on shrinkage compensation of the SLS process by using the Taguchi method. International Journal of Machine Tools & Manufacture, v.42, p. 1203–1212, 2002.

[35] Sharma, P., Verma, A.; Sidhu, R. K.; Pandey, O. P. Process parameter selection for strontium ferrite sintered magnets using Taguchi L9 orthogonal design. Journal of Materials Processing Technology, p.147-151, 2005.

[36] Barros, N.; Bruns, R.E.; Scarminio, I. S. Otimização e planejamento de Experimentos. Campinas : Editora da Unicamp, 1995, p.291.

[37] Rosa, J. L.; Robin, A.; Silva, M.B.; Baldan, C. A.; Peres, M. P. Electrodeposition of copper on titanium wires: Taguchi Experimental Design Approach. Journal of Materials Processing Technology, p. II8I-II88, 2009.

[38] Malik, P. K., Saha, S.K. Oxidation of direct dyes with hydrogen peroxide using ferrous ion as catalyst. Separation and Purification Technology, v. 31, p. 241-250, 2003.

Chapter 6

Application of 2^K Experimental Design and Response Surface Methodology in the Optimization of the Molar Mass Reduction of Poly (3-Hydroxybutyrate-co-3-Hydroxyvalerate) (PHBHV)

Sérgio Roberto Montoro, Simone de Fátima Medeiros,
Amilton Martins Santos, Messias Borges Silva and Marli Luiza Tebaldi

Additional information is available at the end of the chapter

1. Introduction

Biodegradable and biocompatible polymers have attracted great attention, both from scientific and technological fields. Developing new biodegradable materials to replace petrochemical derivatives is demanding and also a challenge for chemists worldwide as well as to developing more efficient synthesis processes with the aim to reduce their production cost. Besides environmental effects due to plastic wastes, there is great interest in the development of biocompatible and biodegradable materials for biomedical applications or in other biomaterials. Regarding such materials, polyhydroxyalkanoates (PHAs) are good examples among them. They are polyesters of hydroxyalkanoic acid, globally manufactured in industrial scale using microbial biosynthesis deriving from renewable carbon sources, in form of storage materials as shows Figure 1. The accumulation of such polymers in granules shape are found in the cell's cytoplasm, their diameter show a wide range variation, from 0.2-0.5μm and they work as glycogen synthesizers and are stored by mammals [1-3].

1.1. General PHAs properties

Awareness of PHAs physical, chemical and biological properties is important, mainly in regards to the development of controlled release systems, once they directly influence, among other factors, microencapsulation processes selection and drug release mechanisms [4-5].

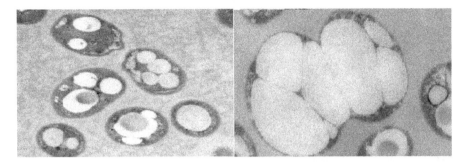

Figure 1. Exponential growth phase containing about 40 - 50% PHB (left) and final accumulation phase containing about 93% PHB (right).

Usually, PHAs are crystalline structures, so their brittleness and low flexibility set limits their application in some biomedical procedures. Thus, lack of superior mechanical proprieties, require modifications, mainly, for medical applications [6].

PHAs molar mass is a crucial factor, once such parameter directly affects mechanical strength, swelling capacity and the ability to undergo hydrolysis, as well as polymers biodegradation rates. Molar mass is related to PHA's crystallinity. Therefore, developing a controlled drug delivery system using PHAs requires an essential preliminary molar mass reduction step. An important factor showing molar mass dependence of biodegradable polymers degradation rates dues to a direct proportionality between this parameter and the polymers Tg, i.e., the lower the polymer molar mass, the lower the T_g value. In addition, degradation rates increasing leads to complex ingredients release rates increasing, having these polymers as polymeric matrixes, they will be promptly absorbed by the body [4]. Moreover, low molar mass PHAs can be used as components on several architecture constructions, such as block and grafted copolymers [7].

1.2. PHB and PHBHV copolymers

Poly(3-hydroxybutirate) (PHB) and its copolymers with hydroxyvalerate (HV) are the most studied PHAs in literature [8]. PHB and PHBHV are completely biodegradable and produced by a variety of bacteria's fermentation, [1] degrading throughout natural biological processes, turning them into excellent candidates for bioactive molecules release system's production [9-15]. Both poly(3-hydroxybutyrate) (PHB) and poly(3-hydroxybutyrate-*co*-3-hydroxyvaler-ate) (PHBHV) are biodegradable thermoplastic polyesters produced by a bacteria known as *Ralstonia eutropha* or *Alcaligenes latus* [16]. PHBHV has emerged as a new generation of PHA in which the surface morphology combined with its lower crystallinity when compared to PHB homopolymers fastens degradation processes. Furthermore, copolymers physical properties can be manipulated by varying the HB and HV's composition [17]. Such advantages make the use of PHBHV copolymers suitable for many applications, once compared to PHB homopolymers. Nevertheless, ISO 10993 highlighted such polymers as safe and non toxic materials indicated to be used in animal's controlled drug release tests [18]. The PHBHV structure is shown in Figure 2.

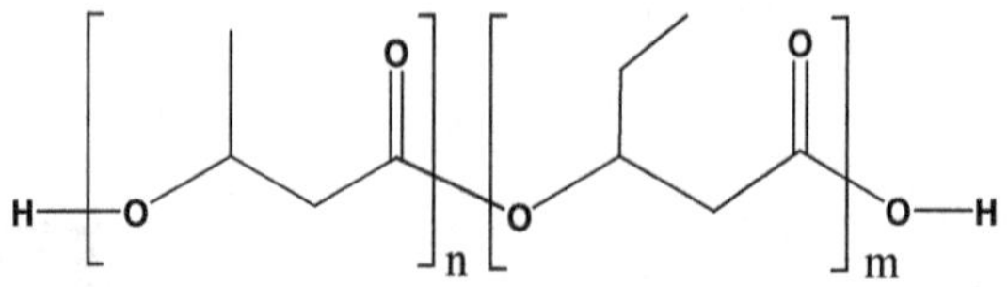

Figure 2. Illustration of PHBHV chemical structure.

Polymers used as carrier agents play a very important role in active ingredients controlled delivery systems, determining, among other factors, drug release rates. In fact, drug release can be performed by different mechanisms: diffusion throught the swelled polymer network, erosion/biodegradation of polymer chains as well as a combination of mechanisms. Basically, the difference between erosion and biodegradation mechanisms is established based on the macromolecule degradation ones. Polymer matrix erosion results, in a macroscopic level, in mass loss, with no modifications on macromolecular units. It can happen in two different ways: either by breaking intermolecular bonds in cross-linked systems, where matrixes are gradually eroded from the external surface, or by main chain's side group hydrolysis, resulting in polymers dissolution, without reducing its molar mass [19].

According to such context, biodegradable polymers are those where chain's breaking results in monomeric units in which molar masses are small enough to be eliminated by normal metabolic pathways. Such breakage can occur by hydrolysis (hydrolytic degradation) or enzymatic attack (enzymatic degradation). Erosive or degradable systems, particularly those that degrade in biological mediums (biodegradable) have found great utility in controlled release systems development.

Polymers natural elimination, after total drug release, is an advantage once it avoids the inconvenience of surgical interventions to remove it. Thus, micro and nano-structured as well as biodegradable and biocompatible systems (microspheres, microcapsules, nano-capsules, nano-spheres) have been developed for drug controlled release [20].

The role that polymers play during the formulation (modulation) of these controlled release systems is very different if compared to an inert conventional excipient for pharmaceutical formulations. Polymers influence not only drug kinetics release, as expected, but also the drug 's stability, toxicity and the compatibility between biopolymers and living organism.

Innumerous techniques have been proposed in literature, regarding polymeric micro and nanostructured materials for drug controlled delivery systems preparation [21].

The most common method used for systems based on PHAs, such as PHB and PHBHV, is called micellization, i.e., - preparation of small micelles through self-assemblage of amphiphilic chains in water [7]. Inside amphiphilic structures of polymeric micelles, hydrophobic drug molecules are distributed within the hydrophobic cores, whereas the shell keeps a hydration barrier that protects the integrity of each micelle. The use of amphiphilic block copolymers is advantageous because they possess unique physicochemical features such as self-assembly and thermodynamic stability in aqueous solution [22]. In order to get amphiphilic structures, surface changes on the hydrophobic segment with a hydrophilic, which is able to stabilize particles, nontoxic and blood compatible materials essential to avoid macrophages recognition, prolonging blood circulation time and sustaining encapsulated drugs release. Many hydrophilic polymers have been suggested for such application. Poly(ethylene glycol) (PEG), for example, is widely used as a hydrophilic nontoxic segment once combined with hydrophobic biodegradable aliphatic polyesters. It was found that incorporated hydrophilic mPEG groups, showing resistance against opsonization and phagocytosis, also presenting prolonged residence time in blood if compared to nanoparticles prepared without mPEG. Nevertheless, it has been demonstrated that surface modifications of a polymer with this nontoxic material reduces side effects risk in comparison to the non-modified polymers [15].

Low molar mass biodegradable block copolymers, in form of amphiphilic micro and nano-particles, were a suggestion of use as sustained release for a variety of hydrophobic drugs [22]. Encapsulate active ingredients in polymeric nanoparticles aims to turn the delivery of effective doses of pharmacologically active substance to a particular site possible, mainly in tumors, for a sustained period of time, avoiding innumerous side effects associated with multiple drug dosing. The defective and leaky structure of tumor vessels and impaired lymphatic system facilitates internalization of polymeric nanoparticles containing drugs, which enhances active agent local effects and protects healthy cells [4].

1.3. Why PHB and PHBHV need modifications?

Besides PHBHV lower cristallinity when compared to other PHAs, its crystalline percentage (55–80%) also needs to be considered. Moreover, it may suffer degradation by conventional melt processing, which limits its use in many specific applications [15]. In order to minimize such problems, PHB and PHBHV are often blended [13-14, 23-26] or used with a mixture of substances such as natural rubber, it is also used in the preparation of composites, [27] or it can be changed through a number of strategies such as using click chemistry [28-29] or modifying the surface and subsequent graftization in a series of monomers [30-35] as well as along with agents: plasticizers, lubricants, antioxidants, photostabilizers and other miscible polymers [36]. PHBHV were modified with natural rubber producing composites with enhanced mechanical proprieties [27]. Ke *et al.* [33] studied thermal properties and in vitro degradation of amide, amine, and collagen-modified PHBHV films aiming to improve biodegradation rates on cytocompatible biomaterials. Biodegradation rates of modified PHBHV were greater than pure PHBHV, offering an alternative to improve such materials properties.

Linhart and *co-authors* [37] showed that amorphous calcium phosphate (ACP) composites and PHB or PHBHV (PHB-ACP or PHBHV-ACP) would be potential bone substitution materials. As it is known, PHAs intrinsic hydrophobic properties restrict some of their applications *in vivo*. Consequently, these materials could be improved by either chemical modification with functional group's introduction or by modifying the topographic surface. PHBHV surfaces were modified with triarylsulfonium salts upon UV irradiation. The process forms species that abstract hydrogen atoms from the PHBHV surface, generating primary radicals which are able to initiate monomer's polymerization by UV-mediation allowing wettability control of produced films, improving their ability for cellular interaction [32]. Vergnol G. *et al.* [10] described the use of PHAs as stent coatings containing the sirolimus drug. Natural PHBHV, poly(3-hydroxyoctanoate) functionalized with carboxylic groups,$PHO_{75}COOH_{25}$, and diblock copolymer PHBHV-*b*-(lactic acid) were sprayed onto metallic stents. P(HBHV-*b*-LA) as coating, enhanced the drug release profile by limiting sirolimus release. Bilayer systems were proposed, it seemed to be very promising, especially in systems with PHBHV and P(HBHV-*b*-LA). Gracida J. *et al.* [24] studied blends of PHBHV/PHEMA´s degradation by fungal activity using the ASTM method and CO_2 measurements to determine biodegrability. Studies showed that PHBHV/PHEMA blends are biodegradable in a ASTM method context.

As previously mentioned, to tailor PHB thermal and mechanical proprieties, its copolymerization with 3HV is commonly performed. Currently, much research work has been published reporting various methods of obtaining a range of PHBHV copolymers with different 3HV content using different carbon nutrition conditions [1,38]. However, molar mass in biosynthesis is typically high and not suitable for systems of drug controlled release. Moreover, commercial PHBHV shows some disadvantages, such as poor thermal stability and high melting temperature. PHBHV´s thermal degradation temperature is close to 160°C and their melting temperature is around 150°C, resulting in a small processing window [39].

In addition, in PHBHV´s biosynthesis only highly hydrophobic polymers are produced and this is unfavorable to the interaction between a biomaterial surface and cells or some other *in vivo* applications. Such polymers need much more versatile modifications in order to obtain new materials with improved mechanical and thermal properties, besides increasing the hydrophilic character [1].

To achieve this goal, the main procedure aims to perform PHBHV molar mass reduction.

1.3.1. Molar mass reduction and further modifications

In this context, aiming to solve the aforementioned problems, polymers molar mass reduction is a fundamental requirement. Such procedure offers, as its main advantage, the possibility to carry a series of modifications including polymer functionalization with terminal vinyl groups, which are highly reactive, in further modification reactions [40].

Many efforts have been made in order to provide PHB molar mass reduction and to improve its copolymers properties or to prepare the material for further modifications.

PHB and PHBHV molar mass reductions can be obtained by thermal degradation [34,41], acid hydrolysis [4, 42-43], transesterification with glycols [44] reduction with $NaBH_4$ [6, 45] or also *in vivo* esterification with PEG [46].

One promising approach to improve their physical properties, adjusting degradation rates, dues to synthesizing block copolymers using telechelic PHB or PHBHV of low molar mass through chemical routes. These modifications involve reactions with bromide or chloride

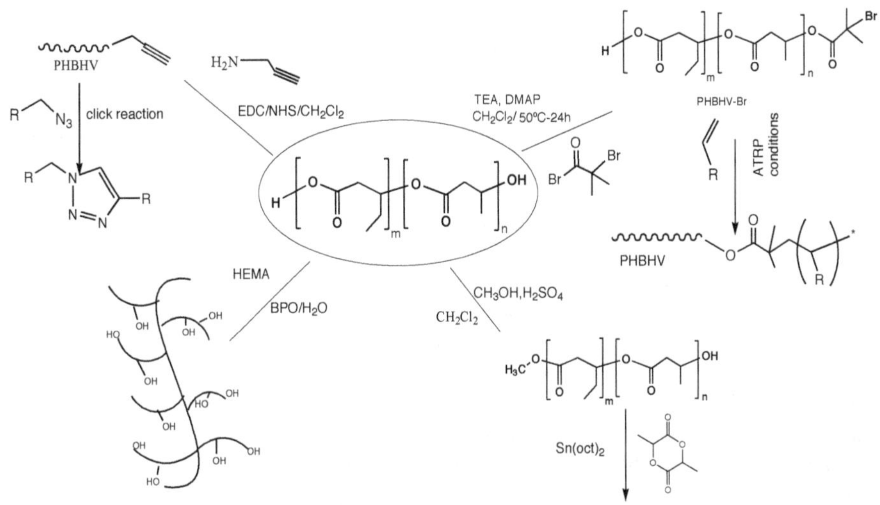

Figure 3. Possible modification reactions in PHBHV with low molar mass.

molecules turning these polymers macro-initiators, which can trigger a new polymerization with various monomers in order to obtain new materials type graft [30-31] or block [12,29,47-48] copolymers with specific properties, for example, amphiphilic systems for drug carrier. Figure 3 shows some possible modifications that can be done to improve PHBHV with low molar mass proprieties.

Arslan H. and *co-authors* prepared lower molar mass PHB-Cl by using a depolymerization process (heating it under reflux with 1,2-dichlorobenzene) and by subsequent choration by means of passing chlorine gas through PHB solutions. The chlorinated PHB (PHB-Cl) were used as macro-initiators in methyl methacrylate (MMA) polymerization aiming to obtain PHB-*g*-PMMA graft copolymers by the atom transfer radical polymerization (ATRP) method [31]. ABA triblock copolymers were prepared trough three consecutive steps. Firstly, natural PHB with high molar mass, was converted into low molar mass PHB-diol by trans-esterification with diethylene glycol. In the next step, PBH-diol pre-polymers were reacted with 2-bromo-2-methylpropionylbromide to obtain PHB-Br macro-initiators which were used to carry out the *tert*-butyl acrylate (*t*BA) polymerization by ATRP. The degradation rate was adjusted according to PHB contents [48]. Spitalský, Z. *et al.* [44] prepared PHB oligomers by alcoholysis using two types of alcohol in the presence of *p*-toluene sulfonic acid as catalyst, which can be used for further crosslinking and chain-extension reactions. Reeve, MS. *et al.* [49] synthesized PHB macro-initiators of low molar mass by methanolysis followed by reactions with AlEt$_3$, which were used to obtain biodegradable diblock copolymers. Hirt, TD. *et al.* [50] obtained telechelic OH- terminated PHB and PHBHV by trans-esterification in order to prepare precursors with reactive end groups which were used to synthesize high-molar mass block copolymers by chain extension. PHB oligomers were prepared and reacted with 2-hydroxyethyl methacrylate

(HEMA) to form macro-monomers with two unsaturated end groups and afterwards graftized with methyl methacrylate to obtain materials that can be used as constituents in acrylic bone cements for use in orthopaedic applications [34]. Oliveira, AM. and *co-authors* synthesized PHBHV-*b*-PNIPAAm block copolymers reacting hydroxyl-caped PHBHV of low molar mass and carboxyl-caped PNIPAAm obtained via reversible addition-fragmentation chain transfer (RAFT) polymerization. The thermo-responsive particles were loaded with dexametasone acetate (DexAc) and showed drug delivery behaviour dependent of temperature, suggesting that these polymeric micelles can be utilized as drug delivery systems [12]. Baran, ET. *et al.* [45] prepared PHBHV of low molar mass via mechanisms of degradation by sodium borohydride ($NaBH_4$). Nanocapsules of PHBHV of low molar mass was prepared and tested for the entrapment of therapeutically active proteins such as those used in cancer therapy. The studies indicated that the use of low molar mass PHBHV was more favorable in increasing entrapment and entrapment efficiency and enzyme activity [46]. Montoro, SR. *et al.* [42] used the methods of acid hydrolysis and trans-esterification with ethylene and hexyleneglycol and also by reductions with sodium borohydride [4,6] to obtain PHBHV with molar mass reduced in order to develop materials suitable to be used as carriers in active systems. Liu,Q. *et al.* [39] synthesized telechelic PHBHV-diols with various molar mass by trans-esterification with ethylene glycol. The results showed that PHBHV-diol was more stable than original PHBHV and the melt-processing window increase gradually with molar mass decrease.

Lemechko, P. *at al.* prepared dextran-*graft*-PHBHV amphiphilic copolymers using two "grafting onto" methods. In the first one, PHBHV oligomers were reacted with SOCl₂ to obtain chloride terminated PHBHV with subsequent esterification with dextran. In the second method, PHBHV oligomers were functionalized with alkyne end groups and graftized onto functionalized dextran via click chemistry reaction. The presence of reactive groups could be interesting to bind bioactive molecules in order to develop heterofunctional nanoparticles [28]. Babinot, J. *at al.* synthesized amphiphilic diblock copolymers with different PHAs of low molar mass. The authors firstly prepared the PHAs oligomers by thermal treatment (190°C) varying the time of reaction and after that, the oligomers were functionalized with alkine function by click chemistry reaction conducting to graphitization with MeO-PEG [29].

The reduction of PHB's molar mass *in vivo* is another strategy, which was used by Ashby RD. *et al.* [46]. The authors controlled PHB's molar mass by adding PEG segments in the incubation medium. PEG interacts with the cellular biosynthetic system which is responsible for P3HB synthesis and regulates the molar mass. A series of PEGs with different molar mass were added to the *Alcaligenes latus* DSM 1122 incubation medium. Such strategy resulted in products of P3HB-PEG diblock copolymers type with reduced molar mass.

Shah *et al.* [15], synthesized amphiphilic biodegradable core–shell nanoparticles by emulsification–solvent evaporation technique using poly(3-hydroxybutyrate-*co*-3-hydroxyvalerate) or poly(3-hydroxybutyrate-*co*-4hydroxybutyrate) diblock copolymers. Copolymers were coupled to monomethoxy poly(ethylene glycol) (mPEG) via trans-esterification reactions. Nanoparticles were found to be assembled in aqueous solution into an outer hydrophilic shell of mPEG, connected to the interior hydrophobic polyhydroxyalkanoate (PHA) copolymer core. Moreover, the morphological examination, by means of atomic force microscope, revealed the nanoparticle's smooth spherically shape. The average particle sizes and zeta potentials of amphiphilic nanoparticles were in the range of 112–162 nm and -18 to -27 mV, respectively. Finally, a hydrophobic drug (thymoquinone) was encapsulated in the nanoparticles and its release kinetics was studied.

In this context, this section presents a study on optimization of ideal PHBHV molar mass reduction due to temperature changes and in concentration of the reducing agent ($NaBH_4$). From a statistical experimental design (2^k Full Factorial) and Response Surface Methodology (RSM), it was determined which of these variables had a greater influence in reducing the molar mass of PHBHV.

Actually, these PHAs are produced in Brazil (pilot-scale), and are considered one of the most promising alternatives due to its properties and low cost.

2. 2^k factorial design

Factorial designs are often used in experiments involving several factors that demand the study of their total effects over a certain response. However, special cases regarding factorial design - in general - are important due to the fact that they are widely used by researchers and because they represent the basis for other considerably valorous planning.

K factor's case is the most important one among them all. Each one presents only two levels. Such levels may be quantitative - two temperature, pressure or time values, etc. - or qualitative - two machines, two operators, a factor's "high" and "low" level-, or, yet the presence and absence of a factor. A complete replication of planning requires 2 x 2 x... x 2 = 2^k observations and is known as 2^k factorial design.

2^k design is particularly useful in some experiment's early stages, when many factors are, probably, observed. It provides a lower amount of turns in which k factors can be studied - by means of a complete factorial design - once there are only two levels of each factor. We must assume that the response is basically linear, considering the chosen factor's ranges [51-54].

2.1. 2^2 design

2^k factorial design simplest type is the 2^2 – two factors, A and B, each one of them holding two levels. We usually think about such levels as "low" and "high" values. Figure 4 illustrates the 2^2 planning. Note that "plans" can be geometrically represented as squares in which $2^2 = 4$ runs - or treatment combinations, forming the square's vertices. Regarding 2^2 planning, it is usual to highlight A and B factors as "low" and "high" levels, using (-) and (+) signs to demonstrate them, respectively. Sometimes it is called "geometric concept for planning".

A special concept is used to underline treatment combinations. In general, a treatment combination is represented by a series of lowercase fonts. If a certain font is shown, the corresponding factor is ran at that treatment combination's high level; if it is absent, the factor is ran at its low level. For example, treatment combination (a) indicates that factor A is at high level and that factor B is at low level. The treatment combination using both factors at low level is represented by (1). Such notation will be used throughout the whole 2^k design's series. For example, a 2^4 treatment, having A and C at high level, as well as B and D at low level, is highlighted by (*ac*) [51,52].

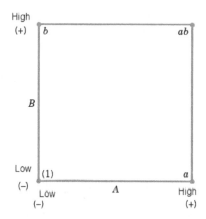

Treatment combination	A	B
(1)	-	-
a	+	-
b	-	+
ab	+	+

Figure 4. Treatment combinations in a 2^2 design [*adapted* 51]

The interest in 2^2 factorial design's effects regards A and B effects, as well as AB second order's interaction factor. According to cases in which fonts (1), *a*, *b*, *ab* are the total of all (*n*) observations performed over these planning points. It is easy to estimate such factor's effects. In order to estimate A factor's main effect, it is necessary to find the observation's average on the right side of the square (Figure 4), - having A at high level – and subtract such average from the observation's average on the left side of the square, where A is at low level, or:

A factor's main effect: 2^2 Factorial design

$$A = \bar{y}_{A+} - \bar{y}_{A-} = \frac{a+ab}{2n} - \frac{b+(1)}{2n} = \frac{1}{2n}\left[a+ab-b-(1)\right] \tag{1}$$

Similarly, B's main effect is found taking the comments' average at the top of the square, being - having B at high level -, and subtract the observations' average at the bottom of the square - having B at low level:

B factor's main effect: 2^2 Factorial design

$$B = \bar{y}_{B+} - \bar{y}_{B-} = \frac{b+ab}{2n} - \frac{a+(1)}{2n} = \frac{1}{2n}\left[b+ab-a-(1)\right] \tag{2}$$

Finally, AB interaction is estimated by finding the difference from diagonal's averages seen on Figure 4, or:

AB effect's interaction: 2^2 Factorial design

$$AB = \frac{ab + (1)}{2n} - \frac{a + b}{2n} = \frac{1}{2n}\left[ab + (1) - a - b\right] \tag{3}$$

The equations quantities in brackets (1), (2) and (3) are called contrasts. For example, A contrast is: Contrast$_A$ = a + ab - b - (1).

According to such equations, the contrasts' coefficients are always (+1) or (-1). A plus (+) and minus (-) signs table, such as on Table 1, can be used to determine the sign of each treatment combination for a particular contrast. The columns names on Table 1 are A and B main effects, AB interaction and I - representing the total. The lines names are treatment combinations. Note that signs on the AB column are the product of A and B columns. In order to generate the contrast from this table, it is demanding to multiply the signals from the appropriate column on Table 1 by the treatment combinations listed on the lines and add. For example, contrast$_{AB}$ = [(1)] + [a] + [b] + [ab] = ab + (1) - a - b [51,52].

Treatment	Factorial Effects			
Combination	I	A	B	AB
(1)	+	-	-	+
a	+	+	-	-
b	+	-	+	-
ab	+	+	+	+

Table 1. Algebraic signs for calculating effects in 2^2 Design.

Contrasts are used to calculate effects estimations and the squares sums for A, B and AB interaction. Regarding any 2^k design with (n) replications, effects estimations are calculated from:

Relation between a contrast and an effect

$$Effect = \frac{Contrast}{n2^{k-1}} \tag{4}$$

And, the sum of any effect's square is:

Any effect's square sum

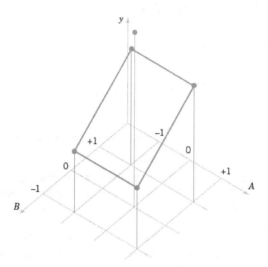

Figure 5. Design with central points [51].

$$SS = \frac{\left(Contrast\right)^2}{n2^k} \tag{5}$$

There is a level of freedom associated to each effect (two levels minus one), so that the error's mean square to each effect is equal to the sum of the squares. The variance analysis is completed by calculating the total sum of squares SS_T (with 4n - 1 level of freedom), as usual, and by getting the squares' error sum SS_E (with (4n - 1) levels of freedom) by subtraction means [51,52].

2.2. Adding central points to 2^k designs

A potential concern in the use of two levels factorial designs is the linearity assumption of linearity in factors effects. Naturally, perfect linearity is unnecessary and the 2^k system will work out well, even when linearity's assumption is approximately kept. However, there is a method to replicate certain points in the 2^k factor. It avoids bending as well as allows estimation, regardless the errors that can be obtained. The method consists in adding points to the 2^k central planning. They consist on replicas races $n_c = 0$ at point x_i ($i = 1, 2,..., k$). An important reason to add replicated races to the planning's center, due to the fact that the central point do not affect usual estimations regarding 2^k planning effects. We consider k factors as quantitative ones. Aiming to illustrate the approach, it was considered a 2^2 plan, with one observation on each one of the factorial points (-,-), (+,-), (-,+) and (+,+) and (n_c) observations on the central points (0,0). Figure 5 illustrates the situation. $\bar{y}_F$ is the average of four runs on the four factorial points and $\bar{y}_C$ is the average of n_c runs at the midpoint [51,52]

If the difference $\bar{y}_F - \bar{y}_C$ is small, the central point will be at or near the flat plane passing through the factorial points and, therefore, there will be no quadratic curve. On the other hand, if $\bar{y}_F - \bar{y}_C$ is large, then a quadratic curve will be present. Squares sums –with an unique freedom degree – to a curve is given by:

Sum of squares sums for curves

$$SS_{Pure\_quadratic} = \frac{n_F n_C \left(\bar{y}_F - \bar{y}_C \right)^2}{n_F + n_C} = \left(\frac{\bar{y}_F - \bar{y}_C}{\sqrt{\dfrac{1}{n_F} + \dfrac{1}{n_C}}} \right)^2 \tag{6}$$

when, in general, n_F is the amount of factorial design points. Such quantity can be compared to error's mean square error in order to test the curve. Note that when the Equation (6) is divided by $\hat{\sigma}^2 = MS_E$ the result will be similar to t-statistic's square, used to compare two means. To be more specific, when points are added to the center of 2^k design, the model that can be found is:

$$Y = \beta_0 + \sum_{j=1}^{k} \beta_j x_j + \sum_{i<j} \sum \beta_{ij} x_i x_j + \sum_{j=1}^{k} \beta_{jj} x_j^2 + \varepsilon$$

once β_{jj} are the pure quadratic effects.

Such squares sum may be incorporated to ANOVA and may be compared to error's mean square, aiming to test pure quadratic curves. When points are added to the center of the 2^k design the matrix for curve (using equation 6) actually tests the hypotheses:

$$H_0 : \sum_{j=1}^{k} \beta_{jj} = 0$$

$$H_1 : \sum_{j=1}^{k} \beta_{jj} \neq 0$$

Furthermore, if the factorial design points are not replicated, n_C central point can be used in order to find an error estimation with $n_C - 1$ level of freedom. A t-test can also be used to test curves [51,52].

3. Response surface planning and methods (RSM)

Response surface methodology, or RSM, is a collection of mathematical and statistical techniques that are useful for modeling and analyzing applications in which the interest response is influenced by several variables and the target is to optimize this response. For example, think of a chemical engineer willing to find temperature degrees (x_1) and pressure

(x_2) to maximize a process' performance (y). The process performance is a function between temperature degrees and pressure, such as in:

$$Y = f(x_1, x_2) + \varepsilon$$

where ε represents the noise or the observed error in the response Y. If we denote the expected response by $E(Y) = f(x_1, x_2) = \eta$, then the surface represented by $\eta = f(x_1, x_2)$ is called "response surface".

We usually, graphically, represent "response surface" as shown on Figure 6, where η is plotted against the x_1 and x_2 levels. Such response surface plots were seen represented on a surface graph in a three dimensional environ.

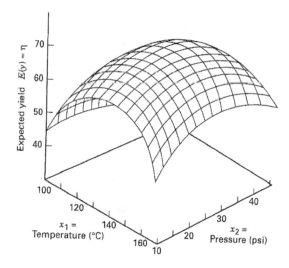

Figure 6. A three-dimensional response surface showing the expected performance (η) as a function between temperature (x_1) and pressure (x_2) [52].

Aiming to observe a response surface plot - particularly in the chapters on factorial designs - to help visualizing its shape, we often plot the response surface's contour, as shown on Figure 7. Regarding the contour's plot, constant response lines are drawn in the x_1, x_2 plane. Each contour corresponds to a response surface particular height. We have seen the utility of contours plots already.

In most of RSM problems, the relation between response and independent variables is unknown. Thus, RSM's first step means finding an adequate rapprochement to the real relationship between Y and the independent variables. Generally, a polynomial of low degree is applied to some independent variables areas. If the answer is well modeled by independent variables linear functions, then the rapprochement function will be the first-order model:

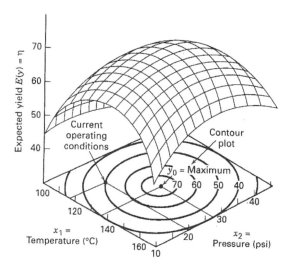

Figure 7. A contour plot of a response surface [52].

$$Y = \beta_0 + \beta_1 x_1 + \beta_2 x_2 + ... + \beta_k x_k + \varepsilon \tag{7}$$

If there is curve in the system, then a higher degree polynomial must be used, such as in the second-order model:

$$Y = \beta_0 + \sum_{i=1}^{k} \beta_i x_i + \sum_{i=1}^{k} \beta_{ii} x_i^2 + \sum\sum_{i<j} \beta_{ij} x_i x_j + \varepsilon \tag{8}$$

Almost all RSM problems use one or both of these models. Of course, it is not feasible that a polynomial model could be a reasonable rapprochement to a real functional relationship regarding the whole independent variables environ. However, regarding a relatively small area, they usually work out quite well.

The minimum squares method is used to estimate parameters to polynomial rapprochements. Response surface analysis is then performed in terms of adjusted surfaces. If the adjusted surface is an adequate rapprochement of the response's true functions, the adjusted surfaces will be almost equivalent to the real system analysis. The model parameters can be estimated most effectively if proper experimental design is used in order to collect data. A design for adjusted response surface is called response surface design [51,52].

RSM is a sequential procedure. Often, whenever we are located on a point on a response surface far from optimum - such as the current operating conditions in Figure 8 -, there is little curving

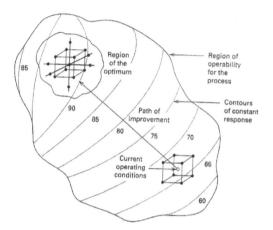

Figure 8. The sequential nature of RSM [52].

in the system and the first-order model will be appropriate. Our goal here is to fast and efficiently lead the experimentalist to optimum's surroundings. Once optimum's region has been found, a more elaborate model - such as the second-order model -, may be applied, and an analysis may be performed to locate the optimum. On Figure 8, we see that the analysis of a response surface can be seen as "climbing a hill" - the top of the hill represents the point of maximum response. If a real optimum is a point of minimum response, then we may think of it as "going down a valley".

RSM further goal dues to determine optimum operating conditions for systems far from the real optimum or to determine an area in the factorial environ, in which operating requirements are fulfilled. Also note that the word "optimum" in RSM is used in a particular sense. "Climbing a hill" RSM procedures aiming to ensure convergence only to an "optimum" place [51,52].

More extensive RSM presentations can be found in Myers and Montgomery (2002), Khuri and Cornell (1996), and Box and Draper (1987). And review paper by Myers et al. (2004) is also a useful reference.

3.1. Steepest ascent method

Frequently, optimal operating conditions initial estimation to the system will be far from real optimum. In such circumstances, the experimentalist's goal is to rapidly move to optimum's general surroundings. We wish to use a simple and economically efficient experimental procedure. When we are away from the optimum, we usually assume that a first-order model is an adequate rapprochement to the true surface in X's small region [51,52].

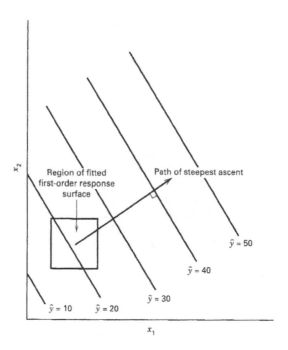

Figure 9. First-order response surface and path of steepest ascent [51,52].

The steepest ascent method is a procedure to sequentially move towards a maximum increase in the response. Of course, if minimization is desired, then we call the technique "steepest descent" method. The adjusted first-order model is:

$$\hat{y} = \hat{\beta}_0 + \sum_{i=1}^{k} \hat{\beta}_i x_i \qquad (9)$$

and the first-order response surface, that is, $\hat{y}$ contours, a series of parallel lines such as those shown on Figure 9. Steepest ascent direction is the one in which $\hat{y}$ rapidly increases. The direction is normal due to adjusted response surface contours. We usually take the line that passes through the center of the interest area and that also is normal to adjusted surface contours as the steepest ascent path. So, steps taken along the path are proportional to the regression coefficients $\hat{\beta}_i$. The real step size is determined by the experimentalist, based on process knowledge or other practical considerations.

Experiments are done throughout the steepest ascent path, until no more increase is observed in the response. Then a new first-order model can be adjusted, a new direction to the steepest ascent is determined, and the procedure continues. Eventually, the experimentalist will reach optimum's surroundings. It is usually indicated by first-order model's lack of adjustment. At this point, additional experiments are performed in order to obtain a more precise optimum estimation [51,52]

4. Experimental methods

4.1. Solvents and reagents

PHBHV (Biocycle®) was the PHA used in the current work, containing 6% HV. The used polymer had a weight average (Mw) and number average (Mn) molar masses of 294.275 and 198.168, respectively, polydispersity index (PI) of the 1,48, melting temperature of 164°C and crystallinity of 55%. The reducing agent used was sodium borohydride, $NaBH_4$ (Colleman) with 97% purity. Solvent chloroform was used as PA (purity > 99%) and to polymer purification it was used methyl alcohol PA (Anidrol) (~ 96% purity). All solvents and other chemicals were used without prior purification by presenting analytical purity.

4.2. General methodology to reduce PHBHV molar mass

Reducing PHBHV molar mass was done by $NaBH_4$ reduction. The methodology used in the procedure was described in Montoro's, SR *et al.* work [4,6,55]

4.3. Statistical design

To optimize the conditions of the current process, we used an experimental design, which included a 2^2 factorial design, with high (+) and low (-) levels, three central points (average) (Table 2), resulting in seven experiments (Table 3).

Factors	Low Level (-)	Medium level (0)	High level (+)
A: $NaBH_4$ concentration (%)	2	4	6
B: Temperature (°C)	50	52,5	55

Table 2. Factors and their respective control levels.

Molar mass and the polydispersity index (PI) were determined by means of Gel Permeation Chromatography (GPC) in a Waters Breeze System equipment.

5. Results and analysis

The experimental matrix for the factorial design is illustrated in Table 4. It is noteworthy that the experiments were performed randomly and an error experimental design was obtained

Experiment number	Factors	
	A	B
1	-	-
2	+	-
3	-	+
4	+	+
5	0	0
6	0	0
7	0	0

Table 3. Parameters used in the 2^k Full Factorial method.

through the mean and standard deviations on repeated central points. The use of factorial design and statistical analysis allowed expressing effectiveness of PHBHV molar mass reduction in molar mass as a linear and quadratic response that can be described as a function with significant variables.

Experiment	% NaBH$_4$	Temperature (°C)	Mn (Da)	Mw (Da)	PI
1	2	50	5804	7576	1,31
2	6	50	4269	4766	1,11
3	2	55	4356	6267	1,44
4	6	55	3848	4052	1,05
5	4	52,5	4521	5190	1,15
6	4	52,5	4545	5244	1,15
7	4	52,5	2994	3920	1,30

Table 4. An experimental matrix for factorial design.

According to results expressed in Table 4 and using Statistica software, there was values regarding parameters for each effect were found (% of NaBH$_4$ and temperature) due to the effectiveness of PHBHV molar mass (M$_n$ and M$_w$) reduction process. We conducted an analysis parameters influence, based on two responses. Therefore, it was necessary to determine which parameters influence really showed statistical significance at a 95% level, it can be observed using a Pareto diagram (Figures 10 and 11).

It can be seen that NaBH$_4$ concentration and temperature variables were statistically significant in response M$_n$ (Figure 11). However, response M$_w$ (Figure 10) showed that NaBH$_4$ percentage showed higher significance if compared to temperature, therefore, both parameters reached significant ranges of statistical significance, adopting 95%. It was observed that the interaction

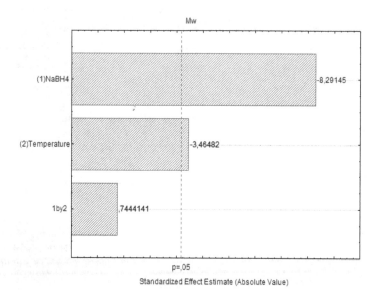

Figure 10. Pareto diagram by M_w.

between $NaBH_4$ variables and temperature showed no importance the PHBHV molar mass reduction process.

Data from factorial design also underwent variance and regression analysis as well as F_0 testing. It has been found - according to data presented on Table 5 - that a PHBHV molar mass reduction model presents coefficients (P-value) and satisfactory regression statistically significant at 95% confidence.

The use of RSM allows the investigation of two variables simultaneously [54] determining regions and molar mass maximum reduction. Figures 12 and 13 show, respectively, the response surface regarding M_w and M_n results, all obtained in PHBHV molar mass reduction experiments, using $NaBH_4$ as the reducing agent.

Results obtained by the experiments have confirmed $NaBH_4$ efficiency, both in reducing PHBHV molar mass and in the uniformity of splitting polymer chains, thereby generating low polydispersity index (PI) values - as presented in Table 4. Montoro *et al.* [6] also showed $NaBH_4$ effectiveness, in both in reducing PHBHV molar mass of PHBHV and the uniformity of splitting polymer chains, if compared to other molar mass reduction means, such as acid hydrolysis and trans-esterification with catalyzed glycols acid.

It was observed that $NaBH_4$ temperature and concentration parameters have strong influence on PHBHV molar mass reduction processes. The response surface analysis gotten from results (Figures 12 and 13) revealed by $NaBH_4$ temperature and concentration higher levels showed

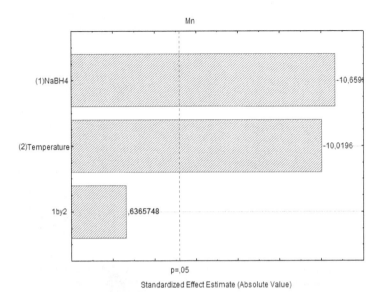

Figure 11. Pareto diagram by M_n.

	Source of Variation	Sum of Squares	Levels of Freedom	Mean Square	F_0	P-Value
M_n	(1) NaBH$_4$	2098152	1	2098152	113,6328	0,001765
	(2) Temperature	1853682	1	1853682	100,3927	0,002116
	1 X 2	7482	1	7482	0,4052	0,569649
	Error	55393	3	18464		
	SS TOTAL	4014710	6			

	Source of Variation	Sum of Squares	Levels of Freedom	Mean Square	F_0	P-Value
M_w	(1) NaBH$_4$	6648662	1	6648662	68,74822	0,003675
	(2) Temperature	1161006	1	1161006	12,00499	0,040498
	1 X 2	53592	1	53592	0,55415	0,510633
	Error	290131	3	96710		
	SS TOTAL	8153392	6			

Table 5. Analysis on variances for table 4 experiment.

greater reduction in PHBHV molar mass. In the current study's specific case, it was found that molar mass reduction optimization occurred in a PHBHV reaction under a 55°C temperature, using 6% NaBH$_4$.

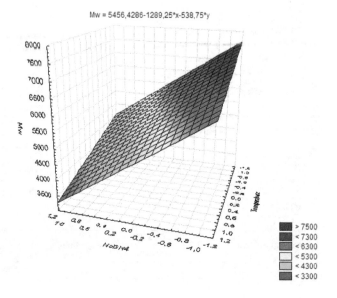

Figure 12. A three-dimensional response surface showing PHBHV molar mass reduction (M_w) as a $NaBH_4$ and temperature function.

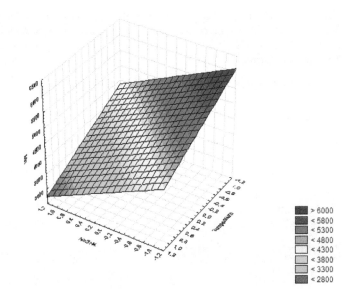

Figure 13. A three-dimensional response surface showing PHBHV molar mass reduction (M_n) as a NaBH$_4$ and temperature function.

6. Conclusion

Factorial designs are often used in experiments involving several factors where it is necessary to study the joint effect of factors on a response. However, several special cases of factorial design, in general, are important because they are widely used in research and because they form the basis for other considerable practical value's plans.

The 2^k design is particularly useful in experimental work early stages, when many factors are probably investigated. It provides the lowest number of runs in which k factors can be studied due to a complete factorial design. There are only two levels of each factor, we have to assume that the response is approximately linear in the range of chosen factors levels.

The use of RSM allows the investigation of two variables, simultaneously, thus determining regions and the maximum reduction in PHBHV molar mass.

The results obtained from the experiments have confirmed the efficiency of $NaBH_4$, both in reducing PHBHV molar mass of and in the uniformity of splitting polymer chains, thereby generating low polydispersity index (PI) values.

It was observed that $NaBH_4$ temperature and concentration parameters have strong influence on PHBHV molar mass reduction processes. The analysis of the response surface gotten from final outcomes revealed $NaBH_4$ temperature and concentration high levels, it increases PHBHV molar mass reduction.

Based in the 2^2 factorial design, ANOVA and RSM techniques, all optimized to reduce PHBHV molar mass. Regression models - the 95% confidence limit – explained data variation (P-Value) for M_w and M_n values. Regarding the response surface analysis, it was found that PHBHV molar mass reduction optimization happened in higher levels of temperature (55°C) and $NaBH_4$ concentration (6%). In general, increasing levels of temperature and $NaBH_4$ concentration resulted in major reductions of PHBHV molar mass.

Acknowledgements

The authors thank CNPq and FAPESP (Process 141794/2010-0 and 2010/15680-9, respectively). We thank the student Ana Valéria Freitas Maia for your collaboration in this work.

Author details

Sérgio Roberto Montoro[1,2*], Simone de Fátima Medeiros[1], Amilton Martins Santos[1], Messias Borges Silva[1] and Marli Luiza Tebaldi[3]

*Address all correspondence to: srmontoro@dequi.eel.usp.br ; montoro.sergio@gmail.com

1 Department of Chemical Engineering, Lorena Engineering School - University of São Paulo, Lorena, São Paulo, Brasil

2 UNESP - Univ Estadual Paulista, Guaratinguetá Engineering School, Department of Materials and Technology, Guaratinguetá, São Paulo, Brasil

3 Federal University of Itajubá-UNIFEI, Advanced Campus of Itabira, Minas Gerais, Brasil

References

[1] Salehizadeh H, Van Loosdrecht MCM. Production of polyhydroxyalkanoates by mixed culture: recent trends and biotechnological importance. Biotechnology Advances 2004; 22, 261-279.

[2] Sudesh K, Abe H, Doi Y. Synthesis, structure and properties of polyhydroxy alkanoates:biological polyesters. Progress in Polymer Science 2000; 25,1503-1555.

[3] Chandra R, Rustgi R. Biodegradable Polymers Progress in Polymer Science 1998; 23:1273–1335.

[4] Montoro S. Redução da massa molar do poli(3-hidroxibutirato-co-3-hidroxivalerato) (PHBHV) para sua posterior utilização no desenvolvimento de sistemas de liberação controlada. PhD Thesis. Engineering School of Lorena 2005.

[5] Terada M, Marchessault M H. Determination of solubility parameters for poly(3-hydroxyalkanoates). International Journal of Biological Macromolecules 1999; 25(1-3) 207–215.

[6] Montoro S R, Sordi M L T, Santos A M, Ré M I. Estudo Cinético da Redução da Massa Molar do Poli(3-Hidroxibutirato-co-3-Hidroxivalerato) (PHBHV). Polímeros: Ciência e Tecnologia 2010; 20(1) 19-24.

[7] Riess G. Micellization of block copolymers. Progress and Polymer Science 2003; 25(8) 1107–1170.

[8] Vidhate S, Innocentini-Mei L, Souza N A. Mechanical and Electrical Multifunctional Poly(3-hydroxybutyrate-co-3-hydroxyvalerate)-Multiwall Carbon Nanotubes Nanocomposites, Polymer Engineering and Science 2012; 52(6)1367–1374.

[9] Grillo R, Pereira AES, Melo NFS, Porto RM, Feitosa LO, Tonello PS, Dias NLF, Rosa AH, Lima R, Fraceto LF. Controlled release system for ametryn using polymer microspheres: Preparation, characterization and release kinetics in water. Journal of Hazardous Materials 2011 186, 1645-1651.

[10] Vergnol G, Sow H, Renard E, Haroun F, Langlois V. Multilayer approach for tuning the drug delivery from poly(3-hydroxyalkanaoate)s coatings. Reactive & Functional Polymers 2012; 72, 260-267.

[11] Kurth N, Renard E, Brachet F, Robic D, Guerin Ph, Bourbouze R. Poly(3-hydroxyoctanoate) containing pendant carboxylic groups for the preparation of nanoparticles aimed at drug transport and release. Polymer 2002; 43, 1095-1101.

[12] Oliveira AM, Oliveira PC, Santos AM, Zanin MHA, Ré MI. Synthesis And Characterization of Thermo-Responsive Particles of Poly (Hydroxybutirate-co-Hydroxyvalerate)-b-Poly(N-Isopropylacrylamide). Brazilian Journal of Physics 2009; 39, 217-222.

[13] Simioni AR, Vaccari C, Re MI, Tedesco AC. PHBHV/PCL microspheres as biodegradable drug delivery systems (DDS) for photodynamic therapy (PDT). Journal of Materials Science 2008; 43(2) 580-585.

[14] Poletto FS, Jäger E, Ré MI, Guterres SS, Pohlmann AR. Rate-modulating PHBHV/PCL microparticles containing weak acid model drugs. International Journal of Pharmaceutics 2007; 345(1) 70-80.

[15] Shah M, Naseer M I, Choi M H, Kim M O, Yoon S C. Amphiphilic PHA–mPEG copolymeric nanocontainers for drug delivery:Preparation, characterization and in vitro evaluation. International Journal of Pharmaceutics 2010; 400,165 -175.

[16] John L, Foster R. Biosynthesis, properties and potential of natural–synthetic hybrids of polyhydroxyalkanoates and polyethylene glycols. Applied Microbiology and Biotechnology 2007; 75,1241-1247.

[17] Pouton C W, Akhtar S. Biosynthetic polyhydroxyalkanoates and their potential in drug delivery. Advanced Drug Delivery Reviews 1996; 28(1)133 -162.

[18] Pötter M, Steinbüchel A. Poly(3-hydroxybutyrate) granule-associated proteins: impacts on poly(3-hydroxybutyrate) synthesis and degradation. Biomacromolecules 2005; 6(2) 552-560.

[19] Porte H, Couarraze G. Microencapsulation processes for the manufacture of systems providing modified release of the active constituent. In: Chulia D, Delail M, Pourcelot Y. (eds) Powder technology and pharmaceutical processes. Elsevier Science, Amsterdam 1994; 513-543.

[20] Benoit M A. Biodegradable Microspheres: Advances in Production Technology, In: Benita S. (ed) Microencapsulation: Methods and Industrial Applications. Marcel Dekker Inc., NY 1996; 35-72.

[21] Poletto F S, Fiel L A,Donida B, Ré M I, Guterres S S, Pohlmann A R. Controlling the size of poly(hydroxybutyrate-co-hydroxyvalerate) nanoparticles prepared by emulsification–diffusion technique using ethanol as surface agent. Colloids and Surfaces A: Physicochemical and Engineering Aspects 2008; 324(2) 105-112.

[22] Liechty W B, Kryscio D R, Slaughter D V, Peppas N A. Polymers for Drug Delivery Systems. Department of Chemical Engineering, University of Texas, Austin, Texas 78712-1062. Annual Review of Chemical and Biomolecular Engineering 2010; 1,149-173.

[23] Srubar III WV, Wright ZC, Tsui A, Michel AT, Billington SL, Frank CW. Characterizing the effects of ambient aging on the mechanical and physical properties of two commercially available bacterial thermoplastics. Polymer Degradation and Stability 2012, 97, 1922-1929.

[24] Gracida J, Alba J, Cardoso J, Perez-Guevara F. Studies of biodegradation of binary blends of poly(3-hydroxybutyrate-co-3-hydroxyvalerate) (PHBHV) with poly(2-hy-

droxyethyl metacrilate) (PHEMA). Polymer Degradation and Stability 2004; 83, 247-253.

[25] Wang S, Ma P, Wang R, Wang S, Zhang Y, Zhang Y. Mechanical, thermal and degradation properties of poly(d,l-lactide)/poly(hydroxybutyrate-co-hydroxyvalerate)/poly(ethyleneglycol) blend. Polymer Degradation and Stability 2008; 93, 1364-1369.

[26] Shi Q, Chen C, Gao L, Jiao L, Xu H, Guo W. Physical and degradation properties of binary or ternary blends composed of poly(lactic acid), thermoplastic starch and GMA grafted POE. Polymer Degradation and Stability 2011; 96, 175-182.

[27] Coelho JFJ, Góis JR, Fonseca AC, Gil MH. Modification of Poly(3-hydroxybutyrate)-co-Poly(3-hydroxyvalerate) with Natural Rubber. Journal of Applied Polymer Science 2010; 116: 718-726.

[28] Lemechko P, Renard E, Guezennec J, Simon-Colin C, Langlois V. Synthesis of dextran-graft-PHBHV amphiphilic copolymer using click chemistry approach. Reactive & Functional Polymers 2012; 72, 487-494.

[29] Babinot J, Renard E, Langlois V. Controlled Synthesis of Well Defined Poly(3- hydroxyalkanoate)s-based Amphiphilic Diblock Copolymers Using Click Chemistry. Macromolecular Chemistry and Physic 2011; 212, 278-285.

[30] Lao HK, Renard E, Linossier I, Langlois V, Vallée-Rehel K. Modification of Poly(3-hydroxybutyrate-co-3-hydroxyvalerate) Film by Chemical Graft Copolymerization" Biomacromolecules 2007, 8, 416-423.

[31] Arslan H, Yesilyurt N, Hazer B. The Synthesis of Poly(3-hydroxybutyrate)-gpoly(methylmethacrylate) Brush Type Graft Copolymers by Atom Transfer Radical Polymerization Method. Journal of Applied Polymer Science 2007, 106, 1742–1750.

[32] Versace DL, Dubot P, Cenedese P, Lalevée J, Soppera O, Malval JP, Renard E, Langlois V. Natural biopolymer surface of poly(3-hydroxybutyrate-co-3-hydroxyvalerate)-photoinduced modification with triarylsulfonium salts. Green Chemistry 2012,14, 788-798.

[33] Ke Y, Wang Y, Ren L, Wu G, Xue W. Surface modification of PHBV films with different functional groups: Thermal properties and in vitro degradation. Journal of Applied Polymer Science 2010; 118(1) 390-398.

[34] Nguyen S, Marchessault RH. Synthesis and Properties of Graft Copolymers Based on Poly(3-hydroxybutyrate) Macromonomers. Macromolecular Bioscience 2004; 4, 262-268.

[35] Koseva NS, Novakov CP, Rydz J, Kurcok P, Kowalczuk M. Synthesis of aPHB-PEG Brush Co-polymers through ATRP in a Macroinitiator-Macromonomer Feed System and Their Characterization. Designed Monomers & Polymers 2010; 13, 579-595.

[36] Srubar III WV, Wright ZC, Tsui A, Michel AT, Billington SL, Frank CW. Characterizing the effects of ambient aging on the mechanical and physical properties of two

commercially available bacterial thermoplastics. Polymer Degradation and Stability 2012; 97, 1922-1929.

[37] Linhart W, Lehmann W, Siedler M, Peters F, Schilling AF, Schwarz K, Amling M, Rueger JM, Epple M. Composites of amorphous calcium phosphate and poly(hydroxybutyrate) and poly(hydroxybutyrateco-hydroxyvalerate) for bone substitution: assessment of the biocompatibility. Journal of Materials Science 2006; 41, 4806-4813.

[38] Bonartsev AP, Myshkina VL, Nikolaeva DA, Furina EK, Makhina TA, Livshits VA, Boskhomdzhiev AP, Ivanov EA, Iordanskii AL, Bonartseva GA. Biosynthesis, biodegradation, and application of poly(3-hydroxybutyrate) and its copolymers - natural polyesters produced by diazotrophic bacteria. Communicating Current Research and Educational Topics and Trends in Applied Microbiology 2007 (2).

[39] Liu Q, Shyr TW, Tung CH, Liu Z, Zhan G, Zhu M,Deng B. Particular thermal proprieties of poly(3-hydroxybutyrate-co-3-hydroxyvalerate) oligomers. Journal Polymer Researcher 2012; 19(1) 1-9.

[40] Hazer DB, Kiliçay E, Hazer B. Poly(3-hydroxyalkanoate)s: Diversification and biomedical applications A state of the art review. Materials Science and Engineering C 2012; 32, 637–647.

[41] Nguyen S, YU G, Marchessault RH. Thermal degradation of poly(3-hydroxyalkanoates): preparation of well-defined oligomers. Biomacromolecules 2002; 3, 219-224.

[42] Montoro SR, Tebaldi ML, Alves GM, Barboza JCS. Redução da massa molecular e funcionalização do Poli(3-Hidroxibutirato-co-3-Hidroxivalerato) (PHBHV) via hidrólise ácida e transesterificação com glicóis. Polímeros 2011; 21(3) 182-187.

[43] Lauzier C, Revol JF, Debzi EM, Marchessault RH. Hydrolytic degradation of isoled poly(β-hydroxybutyrate) granules. Polymer 1994; 35: 4156-4162.

[44] Spitalský Z, Lacík I, Lathová E, Janigov I, Chodák I. Controlled degradation of polyhydroxybutyrate via alcoholysis with ethylene glycol or glycerol. Polymer Degradation and Stability 2006; 91, 856-861.

[45] Baran ET; Ozer N, Hasirci V. Poly(hydroxybutyrate-co-hydroxyvalerate)nanocapsules as enzyme carriers for cancer therapy: an in vivo study. Journal of Microencapsulation 2002; 19, 363-376.

[46] Ashby RD, Shi F, Gross RA. Use of Poly(ethylene glycol) to Control the End Group Structure and Molecular Weight of Poly(3-hydroxybutyrate) Formed by Alcaligenes latus DSM 1122. Tetrahedron 1997; 53(45)15209-15223.

[47] Zhang X, Yang H, Liu Q, Zheng Y, Xie H, Wang Z, I Cheng R. Synthesis and Characterization of Biodegradable Triblock Copolymers Based on Bacterial Poly[(R)-3-hydroxybutyrate] by Atom Transfer Radical Polymerization. Journal of Polymer Science: Part A: Polymer Chemistry 2005, 43, 4857- 4869.

[48]   Adamus G, Sikorska W, Janeczek H, Kwiecien M, Sobota M, Kowalczuk M. Novel block copolymers of atactic PHB with natural PHA for cardiovascular engineering: Synthesis and characterization. European Polymer Journal 2012; 48, 621-631.

[49]   Reeve MS, McCarthy SP, Gross RA. Preparation and Characterization of (R)-Poly(β-hydroxybutyrate)-Poly(ϵ-caprolactone) and (R)-Poly(3-hydroxybutyrate)-Poly(lactide) Degradable Diblock Copolymers. Macromolecules 1993; 26, 888-894.

[50]   Hirt TD, Neuenschwander P, Suter UW. Telechelic diols from poly[(R)-3-hydroxybutiric acid] and poly{[(R)-3-hydroxybutiric acid]-co-[(R)-3-hydroxyvaleric acid]}. Macromolecular Chemistry and Physics 1996; 197, 1609-1614.

[51]   Montgomery DC, Runger GC. Estatística Aplicada e Probabilidade para Engenheiros, 5ª ed. Rio de Janeiro: LTC; 2012.

[52]   Montgomery DC. Design and Analysis of Experiments, 6ª Ed. EUA: John Wiley & Sons, Inc; 2005.

[53]   Dean AM, Voss DT. Design and Analysis of Experiments. EUA: Springer; 1999.

[54]   Neto BB, Scarminio IS, Bruns RE. Como Fazer Experimentos - Pesquisa e Desenvolvimento na Ciência e na Indústria, 4ª ed. Porto Alegre: Bookman; 2010.

[55]   Montoro SR, Maia AVF, Benini KCCC, Tebaldi ML, Silva MB. 11° CBPol 2011: conference proceedings, October, 16-20, 2011, Campos do Jordão/SP-Brazil.

Permissions

The contributors of this book come from diverse backgrounds, making this book a truly international effort. This book will bring forth new frontiers with its revolutionizing research information and detailed analysis of the nascent developments around the world.

We would like to thank Dr. Messias Borges Silva, for lending his expertise to make the book truly unique. He has played a crucial role in the development of this book. Without his invaluable contribution this book wouldn't have been possible. He has made vital efforts to compile up to date information on the varied aspects of this subject to make this book a valuable addition to the collection of many professionals and students.

This book was conceptualized with the vision of imparting up-to-date information and advanced data in this field. To ensure the same, a matchless editorial board was set up. Every individual on the board went through rigorous rounds of assessment to prove their worth. After which they invested a large part of their time researching and compiling the most relevant data for our readers. Conferences and sessions were held from time to time between the editorial board and the contributing authors to present the data in the most comprehensible form. The editorial team has worked tirelessly to provide valuable and valid information to help people across the globe.

Every chapter published in this book has been scrutinized by our experts. Their significance has been extensively debated. The topics covered herein carry significant findings which will fuel the growth of the discipline. They may even be implemented as practical applications or may be referred to as a beginning point for another development. Chapters in this book were first published by InTech; hereby published with permission under the Creative Commons Attribution License or equivalent.

The editorial board has been involved in producing this book since its inception. They have spent rigorous hours researching and exploring the diverse topics which have resulted in the successful publishing of this book. They have passed on their knowledge of decades through this book. To expedite this challenging task, the publisher supported the team at every step. A small team of assistant editors was also appointed to further simplify the editing procedure and attain best results for the readers.

Our editorial team has been hand-picked from every corner of the world. Their multi-ethnicity adds dynamic inputs to the discussions which result in innovative outcomes. These outcomes are then further discussed with the researchers and contributors who give their valuable feedback and opinion regarding the same. The feedback is then collaborated with the researches and they are edited in a comprehensive manner to aid the understanding of the subject.

Apart from the editorial board, the designing team has also invested a significant amount of their time in understanding the subject and creating the most relevant covers. They scrutinized every image to scout for the most suitable representation of the subject and create an appropriate cover for the book.

The publishing team has been involved in this book since its early stages. They were actively engaged in every process, be it collecting the data, connecting with the contributors or procuring relevant information. The team has been an ardent support to the editorial, designing and production team. Their endless efforts to recruit the best for this project, has resulted in the accomplishment of this book. They are a veteran in the field of academics and their pool of knowledge is as vast as their experience in printing. Their expertise and guidance has proved useful at every step. Their uncompromising quality standards have made this book an exceptional effort. Their encouragement from time to time has been an inspiration for everyone.

The publisher and the editorial board hope that this book will prove to be a valuable piece of knowledge for researchers, students, practitioners and scholars across the globe.

List of Contributors

Helder Jose Celani de Souza
Siemens Healthcare Diagnostics, Sao Paulo State University – UNESP, Production Engineering Department, Guaratingueta, SP, Brazil

Cinthia B. Moyses and Fernando Lopes Alberto
Fleury Diagnostics, São Paulo, SP, Brazil

Fabrício J. Pontes and Ubirajara R. Ferreira
Sao Paulo State University, UNESP, Guaratingueta, SP, Brazil

Roberto N. Duarte
CEFET - São Joao da Boa Vista, SP, Brazil

Carlos Eduardo Sanches da Silva
Federal University of Itajuba, Itajuba, MG, Brazil

H. Teimouri, A. S. Milani and R. Seethaler
School of Engineering, University of British Columbia, Kelowna, Canada

Cristie Diego Pimenta, , Roberto Campos Leoni, Ricardo Batista Penteado, Fabrício Maciel Gomes and Valério Antonio Pamplona Salomon
FEG UNESP – SP, Brazil
Guaratinguetá-SP, Brazil

Rosinei Batista Ribeiro
FATEA – SP / Brazil
Lorena – SP, Brazil

Eduardo Batista de Moraes Barbosa
São Paulo State University "Júlio de Mesquita Filho" – UNESP, School of Engineering at Guaratinguetá – FEG, Guaratinguetá, SP, Brazil

Messias Borges Silva
São Paulo State University "Júlio de Mesquita Filho" – UNESP, School of Engineering at Guaratinguetá – FEG, Guaratinguetá, SP, Brazil
University of São Paulo – USP, School of Engineering at Lorena – EEL, Estrada Municipal do Campinho, s/n – Lorena, SP, Brazil

Ana Paula Barbosa Rodrigues de Freitas and Carla Cristina Almeida Loures
University of São Paulo (USP), Brazil
São Paulo State University (UNESP), Brazil

Marco Aurélio Reis dos Santos, Geisylene Diniz Ricardo and Fernando Augusto Silva Marins
São Paulo State University (UNESP), Brazil

Hilton Túlio Lima dos Santos
University of São Paulo (USP), Brazil

Gisella Lamas Samanamud
University of Texas at San Antonio (UTSA), Brazil

Mateus Souza Amaral
University of São Paulo (USP), Brazil

Leandro Valim de Freitas
Petróleo Brasileiro SA (PETROBRAS), Brazil
São Paulo State University (UNESP), Brazil

Sérgio Roberto Montoro
Department of Chemical Engineering, Lorena Engineering School - University of São Paulo, Lorena, São Paulo, Brazil
UNESP - Univ Estadual Paulista, Guaratinguetá Engineering School, Department of Materials and Technology, Guaratinguetá, São Paulo, Brazil

Simone de Fátima Medeiros and Amilton Martins Santos
Department of Chemical Engineering, Lorena Engineering School - University of São Paulo, Lorena, São Paulo, Brazil

Marli Luiza Tebaldi
Federal University of Itajubá-UNIFEI, Advanced Campus of Itabira, Minas Gerais, Brazil

Printed in the USA
CPSIA information can be obtained
at www.ICGtesting.com
JSHW011327221024
72173JS00003B/81